SCIENTIFIC REVELATIONS

austintorn@aol.com

SCIENTIFIC REVELATIONS

Ongoing scientific investigations discover
What's behind the happenings,
Such as that we can feel other's states
Through empathy via mirror neurons
And that we can even further
Make models of other's minds;

While non-science dogma
Decides its arbitrary information
All at once.

Science may well disagree, through proof,
With some higher mammal's wishes
For how things ought to be.

Science does not then,
Entail "magic", "physic powers", superstition,
Special creation in lieu of evolution,
Wishes, beliefs, interdimensions of ETs,
"Soul", or other such invisible claims;

But, it can explain the reasons
Behind these erroneous modes of thinking,
And, in addition, if say,
That something like brain waves
Could be picked up by another,
Then science would find the natural basis.

THE ETERNAL GROUND-STATE

To be or not to be,
That is the question (or answer)
Of "Why there is something rather than nothing?"
[Seemingly this the most unanswerable question!]

(Zero-point energy, Quantum foam,
Vacuum Energy, etc.)

There can be no level
Of "God" adjoining here,
For this energy can
Neither be created nor destroyed.

It will always be with us and it always was.

This is because it cannot
Have come from Nothing,
Nor can it ever become Nothing.

This cuts out all possible Gods
At the source (one which never was),
Including the Deity.

If there had been a total lack of anything
Then that would still be the case,
But, it is not, for there is something;
So, Nothing wasn't possible.

The "great question" of
Why there is something rather than Nothing
Is stated backwards,
As if Nothing could do anything.

It can't, for it has no properties.

So, *Why does anything exist at all?*
Because it had to,
For it is impossible for there
Not to be something,
As Nothing is unproductive.

The concept of nothing yet plays a role,
For the emitted particles cancel out;
Thus there is no paradox about
The specific and astronomical amount
Of positives and negatives in the universe,
Nor where it springs, since anywhere,
Nor when, since any time.

Nor is God the energy, the particles emitted,
Or Nature, for these states are restrictive.

If one still insists that God is nature,
This is but like saying
That the universe is the cosmos,
For a rose is ever a rose is a rose.

One boxes themselves in, then,
To claim God, for then one
Cannot explain what is beyond the walls,
And thus has only opened the gap wider
By the posing a larger question
Instead of addressing an answer.

As the ground-state must be eternal,
There was no creation, and thus no Creator.

LIFE ELSEWHERE?

Is there a way that we could determine
If life flourishes throughout the universe,
Seeing that we can't really travel to most of it?

Yes, for if we find independent life
Somewhere nearby, say,
On Saturn's moon, Titan,
Then we'd expect
The same everywhere.

PATENTING THE TAO

All is solved with the TAO,
With just a few word-details now
For the lawyers to work on,
Such as with the TAO
That is not the OAT.

I tried to get a copyright on the TAO symbol,
I guess God owns it.

We will still get a PatPending, Prof—
It's in the works...

Note:
The lawyers don't like to say
That the TAO is not a duck,
The TAO is not a dog, etc.,
On up through everything
Real and imaginable.

We will also have a logo of sorts,
Maybe a cube,
With some letters on each face
To show that we've covered everything,
Like, for one face:

G U T
D O A
N E O

Taoist propriety and ethics emphasize
The Three Jewels of the Tao:
Compassion, moderation, and humility,
While Taoist thought generally focuses
On nature, the relationship between humanity,
And the cosmos,
Health and longevity,
And wu wei (action through inaction),
Which is thought to produce harmony
With the universe. —Wiki

(God is not really associated with the TAO)

OK, Prof—I was up late,
Burning some of your midnight lamp oil
At the Supreme Court.

The Good News:
God did not show up to claim the TAO rights.

The Bad News:
The court's lawyer is named Fredrick;
He's the attorney general for the universe.

The Not So Bad News:
I am the PatTorney general,
One letter higher than Fredrick,
Making him a peon.

They said that if the TAO
Couldn't be spoken
Then I didn't know
What I was talking about.

So, I presented the +/- balance
Of Nature's Account
And they liked it.

Some excerpts from my presentation:

From a Western perspective,
The Taoist view of sexuality
Is considerably more at ease.
The body is not viewed as
A dangerous source of evil temptation,
But rather as a positive asset.

Taoism rejects Western mind-body dualism;
Mind and body are not set in contrast
Or opposition with each other.
Sex is treated as a vital component
To romantic love. — Wiki

LIVING IN INTERESTING TIMES

Eighty five years ago our universe
Consisted of a single galaxy, our Milky Way.
Now we know there are over 400 billion galaxies!

The three pillars of the big bang, are
(1) The Hubble Expansion,
(2) The existence of the CMB radiation, and
(3) Observations of the abundance of light elements.

Because of cosmic acceleration,
All galaxies outside of our
Own bound cluster will disappear.
With no galaxies there will be
No tracers of the Hubble expansion.

Interestingly, with no such tracers,
All evidence of the existence of dark energy,
And an accelerating universe will also disappear.

The only period
When dark energy is detectable is now;
At much earlier times
It had a negligible effect
Upon the expansion,
And, at late times, it drives out
All tracers of the expansion.

The CMB becomes unobservable
Even in principle,
As the peak wavelength is driven
To a length larger than the horizon.

The final key bit of evidence for the Big bang
Rests crucially on the fact
That relic abundances of deuterium
Remain observable at the present day...

But, by then, both the QSOs
And the Lyman-Systems
Will have red shifted outside of the horizon.

What will observers of the far future infer then?

Their picture of the universe
Will not be significantly different
Than that which Einstein had
When he developed general relativity:
A static universe in which our galaxy
Was surrounded by eternal empty space.

UNIFICATION OR SEPARATION?

Electricity and magnetism each
Lead to the other, being transformational.
They facilitate action and motion
Through EM's push-pull of regularity.

The strong force binds the atomic nucleus,
Barely beating EM's repelling force.
The weak force counters strong's stability
Through decay that promotes changeability.

Electromagnetism and the weak force
Unify when the temperature gets hot,
As during the Big Bang, and they oppose
The strong force as duality's balance.

What about gravity? Where has it been?
It needs matter and motion to exist
And so it is the blended result of
All the forces, a secondary effect.

Dualities seem to assist nature:
Good/evil, on/off, hot-cold, man/woman,
Up/down, left-right, here-there, past-future,
And, so, none can exist without the other.

There can be no more unification,
For what One could be versatile enough
To form both the electroweak force and the
Strong—as different as the north/south poles.

THE SCIENCE OF CONSCIOUSNESS

(Of Brian Silston)

We can exist in the world
Being both subjectively self aware
And not subjectively self aware,
Depending on circumstances.

For instance,
Brain injury can permanently remove
Subjective self awareness
But the individual can continue to exist.

Anesthesia can remove awareness
For a period of time,
TMS (transcranial magnetic stimulation)
Can do the same.
Drugs can either severely impair
Awareness or heighten it—
Which is to say, the degree of information
That makes its way up
To subjective awareness varies greatly.

So, subjective awareness can be turned on and off,
But subconscious being must always be on.
The subconscious systems in the brain
Interpret and process information,
Then selectively feed relevant information
Up to awareness.

It is possible to be alive
And act without subjective awareness,
And in fact we can jerry-rig a system
Into forming a memory in a specific way.
Hypnosis is just such a thing.
When the memory is recalled,
It may bear little resemblance
To the actual experience or event,
And was in effect placed there
By a hypnotist in such a way
As to have exploited the subconscious system.

We then have a subjective experience
Of a fake memory.

This points out that the subject's interpretation
Of the subject's relationship to objects can
And would be different
Under the effects described above.

This tells a few important things —
Our brains process information
In the absence of subjective self awareness,
Especially in brain areas outside
The cerebral cortex.

This is the case in infants,
Whose subcortical structures develop first.
These areas create the mental model of the world.

Now to form a memory the brain must
Have some basis for the world
Which it does not have straight out of the womb.
The connections simply have not developed yet.

Master glial cells quickly multiply,
Lay out a scaffolding of sorts
And guide neuron development
To the appropriate areas
Where they will form synapses.

Before this is done
There is no subjective self awareness.

It is possible in extreme cases
For a handful of memories
To form very early on,
But these memories have
Been polluted over time,
Like all memories, by recall,
So the memory we think
We formed during this time
Is more of an evolution
Of a very crude experience.

This is necessarily so
Because the brain hasn't worked out
What things are at that point in development
In order to generate the adequate description
Of them at that point in time—
Ex., what a wire is, what lights are,
What faces are, etc., etc.

The next important point
Is that the brain areas
Associated with implicit structures
Matures before the explicit structures.

The consequence of this
Is that the infant
Doesn't subjectively know
That it is learning,
But the brain is working to develop
A crude model of the world in the meantime.

The infant is guided by instinct,
Which we now generally refer to
As implicit memory or the implicit system.

It is mandatory that the implicit,
Subcortical structures mature quickly (primacy)
And are working properly
Because they also control breathing,
Heart rate, digestion, and other things
Which we don't have to actively think about to do.

During this learning the brain is forming
The explicit, declarative system,
Where glial cells are customizing
Neuronal connections,
Myelinating axons,
And nourishing the neurons.

At a critical point,
The brain connects these systems
In such a way to produce subjective awareness.
The crucial point now

Is that the implicit system
Is communicating with the explicit system—
Making the associations we perceive
At the level of subjective awareness.

Intentionality is then determined
By our history according
To these interactions.
In other words,
Intentionality is a form of decision making
At the subjective level
In the context of a self aware being.

A decision will include influences
From the implicit or subconscious systems
And those will be largely undetectable.
Some intentionality is the result of instruction,
For example: I am going to reframe
How I approach my physical pain
Because I heard that positive attitudes
Lead to less severe reports of subjective pain.

This is only a start, for intentionality requires
Further explanation from this perspective
Because it leads to physical changes in the brain.

Meditation is a good example
Of a top down phenomena
That leads to physical changes.

This is still consistent with the theory
That neurons that fire together wire together,
That is, the more a particular thought is thought,
Or memory recalled,
The stronger those connections become,
And this leads to the physical changes we see.

We, like most other animals,
Are mostly automated—
That is, we can act and exist
Without being subjectively
Self aware of the actions.

The subjectively self aware self
Then can be turned on and off
(By drugs, anesthesia, injury, TMS, etc.).

This implies that this type of self consciousness
Is dependent on certain connections
And interactions between physical systems,
Perhaps specific types of cells in a certain order
In the implicit and explicit networks.

ESCHER TAMES INFINITY

But, not really,
For the space at the edge
Gets really compressed.

So, how many sides does a circle have?

Two. The inside and the outside.
We see many grand concepts,
But many can't see beyond
The edge of their nose in real life.

That which is not can inform us, too,
Such as infinity never happening.

It tells us that time had to begin,
And could not have been forever,
At least for de-finite things.

Zero, the most important number,
Tells that there must be symmetry,
Lest a specific number of things
Become "special", which it never can.

THE 2ND GENESIS

For 100 million years
After the birth of the universe,
Space was dark and mostly formless.

That's 100 million years.

No stars.

It was not interesting in the least.

It was mostly hydrogen and helium,
With faint traces of lithium and beryllium.

It was an abysmally black "void";
Darkness was upon the face of the deep.

Who would have bet on this dark horse
Running through a 100 million year night?

Then hydrogen caught fire
And so the stars were born.

In these blast furnaces,
Atomic nuclei were crushed,
Burned,
And transmuted into
More complex elements.

That was the second creation,
The one that really mattered.

We contain those elements.

Parts of those stars
Are in our blood, bones, and skin.

We are those stars.

HELL ON EARTH

Hydrogen sulfide at 800 ppm leads to death,
Yet, paradoxically, we need it to survive.

We can detect it at even 0.0047 ppm;
It is the smell of rotten eggs.

Some 250 million years ago,
The outlook for life on Earth
Was very grim indeed.

The Permian Era
Was unrelentingly harsh
And the single most devastating
Extinction even was underway.

Carbon dioxide emissions
From massive volcanic eruptions
In Siberia had triggered
A chain of environmental changes
That had left oxygen levels
Dangerously low in the world's oceans.

This shift in ocean chemistry was bad news
For oxygen-breathing marine species.

What purposeless fates were
To befall life so "unprecious"?

What sulfur fumes were to arise from the depths?

Anaerobic organisms such as
Green sulfur bacteria
Thrived and flourished
Under the low oxygen conditions.

Their success made the oceans
All the more inhospitable
To its remaining aerobic inhabitants,
Since the bacteria generated
Vast quantities of hydrogen sulfide (H2S).

The lethal gas then diffused into the air,
Wiping out plants and animals on land.

By the end of the Permian Extinction
95% of marine species
And 70% of the terrestrial ones had perished.

The creatures that survived
This Hellish catastrophe
Were the only ones who could tolerate H2S,
And, in certain cases, even consume it;
Thus, we humans have retained
Some affinity for it.

H2S increases the responsiveness
Of neural circuits,
And even protects stressed neurons.

It dilates blood vessels,
Controlling blood pressure
And protecting the heart.
It regulates contractibility
Of smooth muscle cells in the lungs,
And does the same for the small intestine,
Regulating movement of material
Through the gut.

H2S "hibernation" can even
Protect vital organs from damage
Until energy supply levels
Return to normal,
Such as during trauma,
By helping to maintain
A baseline metabolism.

We arose from Hell,
Once upon a time,
It would seem,
But brought a useful
Part of it along.

FROM EUCLID'S STRAIGHT LINE SHINE

(After K.B. Robertson)

To Isaac's golden apple
And its gravitational extension of geodesic grapple.

Isaac's apple tree is still not bare,
That falling apple waking him there
From his silly pursuits of alchemy
And of banishing the holy trinity

Ironically, he worked at Trinity College, unbanned,
The only one there who was not an Anglican.

A catchy little tune
That most anyone can croon,
By the curved silvery lighted rune
Of Albert's expeditionary moon.

This moon was still there,
Even when no one was aware.

Dr. Einstein's waking inspiration
May yet sweep the sleeping nation
At the slightest provocation.

Maybe, but few are going along,
Since the "bed gravity" is too strong.

An unrecognized solution
That could start a peaceful revolution.
Big Bang Gangology's further confirmation
Of their favorite libation -
More denial and debate
With the orphaned Steady State.

The Steady State is of an orphan's fate
Since it is a parentless uncaused state.

Behold Albert's resurrected Lambda smart bomb
With enduring aplomb.
Benevolent bomb
Leaves all the buildings and people intact;
Takes 4-D space-time to get them back on track.

Gravity is the 4th dimension—
That of everything's expansion.

Asked the teacher what gravity was,
An' all he said is what gravity does.
Said I wanna know why,
Not how things fall
Teacher said nobody knows that one at all.

(good rhyme with 'was' and 'does')

Asked the people on the 6 O'Clock news;
They said on that we have no views.
Same thing happened in a physics lesson—
A picture of Newton gave a puzzled expression.

Like Dorian Grey from his fine times,
Newton's theory changed into Einstein's.

Still wanted to know what gravity is,
So I went outside and continued the quiz.

In the darkness I alit from the Wiz,
And tried to make sense of this world of His.
Now I've found the answer to life's dark quiz:
One must live this life by what light there is.

If nothing is, then there is no quiz,
No right or wrong from a testing Wiz,
So I'll just remain the same as His,
Living out this earthly dream, as is.

Asked a mathematician about all encumbered
And he took all day saying gravity is numbers.
So I lit one up and, went into suspension,
Tintanambulating beyond the 3rd dimension.

'Twas a new dimension
Of another extension.

The answer appeared as a gentle kiss,
So I wrote another poem and it goes like this...

A Brief History of Rhyme—
A Celebration of Space-Time amalgamation
Liberated from extended Fragmentation.

Poetic science appliance.
The exclusion of politics from science.
A scientific paradigm shift alliance.

Einstein's presently abanded
Unified Field reinstated w'out mathematics.

$G = 4D$

Scientific 'mainstream' panics
At the joining of Einstein's field
With Planck's quantum mechanics

Infinities arise from the singularity,
But then maybe that is the answer, verily.

Democritus foresaw the invisible atom,
But since then his discovery
Is found with substratum
From antiquity,
And ubiquity
The continuous wave was the rave.

Until they invented the wavicle.

Faraday found the cathode ray.
Thompson uncovered electrons one day.
Rutherford encountered protons a different way.
Understood Maxwell's waves
Beneath the celestial hood,
Discovered electromagnetic fields
As no one thought he would.

The wave emitting electron could not be subdivided—
At first it was whispered, then openly confided.

Yet along came smaller mysterious articles,
Of Max Planck's curiously indivisible 'particles':

(They could never walk the Plank)

Transforming a mystic world of electrostatics
Into a truculent tangle of quantum mechanics.

Conceptual doors were opened
For the entrance of protons,
But no comprehensive vacancy
For the residential photons.

Other atomic tenants varied in weight,
Height and disguise,
But the photon is always
The same null value and size.

An undulating atom might change
Its balance or valence,
While the unchanging photon
Showed no such talents.

Vigils are kept to find it changing its station,
While its stubborn identity confirms
In black body radiation.

(Lucky that photons don't pile up on the floor)

At dollar conventions where no change is invited,
Twenty nickels sit down to an audience excited.

The quantum takes for granted inclusion,
While greenbacks resent the currency intrusion.
To and from spherical shells the electron darts -
While the unchanged quantum
Arrives before it departs.

(Here today; gone yesterday)

If you're looking for a message in here,
It's of Max Planck's quanta
And Niels Bohr's spheres.

Invincible in principle,
Newton's Mechanics are sure as shooting,
While quantum mechanics are robbing and looting.
Evolutionary experiments are eclectic,
But the final conclusions are photo-electric.

(Robbing Peter to Pay Paul)

As though these convulsions are not enough,
Reality panned out some other stuff.
The only certain universal permanency,
Is Einstein's constant light-speed
And Heisenberg's indeterminacy.

(That's for sure: uncertainty)

Einstein's fort was special & general relativity,
While his Nobel Prize was for photo-electricity.
Uncle Albert having firstly proven to be right—
Ahead of Brownian motion and the speed of light.

(The first real verification of the atom.)

This century old issue of size
Is how Einstein won the Nobel Prize—
How the peace loving master-blaster stayed alive,
In the timely year of 1905.

Anaxgoras of pre biblical days
Took big and little to greater heights and stays,
He said "There's always something larger than large,
And always something smaller than small."
Perhaps the smallest-large statement of all.

(He was a small medium at large.)

May this admonition of illusion
Be this brief sonnet's conclusion.

The 20th century path has been historically rough—
To the point of surrendering enough of this stuff.

THE UNIVERSAL FRAMEWORK

Is not all that important,
But for TOE enthusiasts,
Being that it is much older
And thus more basic and primitive
Than the composite complexities
Of us and our planet.

It is also fairly inaccessible directly,
Being so vast and far away.

Even its TOE is not so interesting,
This being some symmetric
And simple material spewed into
An outward movement
From some eventually
Causeless state.

Our existence trumps it
In every way.

THE OPTIMIZED MIND

The mind (always the brain)
Is not just “good enough”,
Or a “kludge” of a contraption,
But is one of the most
Optimal devices we know.

Nor does it strictly use logic,
As its Artificial Intelligence
“Imitators” found out
When trying simple
Mathematical models;
But employs vague
And generalized notions.

Logic can only be “right” or “wrong”,
But, an emotion,
An inseparable part of cognition,
Cannot be, but we can feel them
As “good” or “bad”.

While evolution is a “tinkerer”,
It was cleverer than
We might think.

We neither have a single,
General purpose learning system
Nor a myriad of special-purpose systems.

There are 4-5 core systems
In human babies and nonhuman animals
For representing:
Objects, actions, number, space,
And possibly social partners.

When the language module
Is added, we can manipulate
And combine the core systems
In various ways that lead
To high level cognition.

Language and cognition
Co-evolved jointly,
Now a marvel of optimality.

Now, what is it about logic
That misguided all of the
Early artificial intelligence research?

It is that exact matches
Can almost never be achieved,
And so our memories
Must be necessarily
Vague and uncertain,
But, they can go from vague
To crisp representations.

Close your eyes
And imagine a familiar object.
It is vague compared to a perception
Of the object with eyes wide open.

During perception, “bottom up” signals
From the visual system
Interact unconsciously with
The vague “top down” signals
Coming from memories.

Only when perception runs its course
Does perception become conscious.

But for this, we would not be able
To understand anything.

The mind-brain has had billions of years
To evolve optimally to solve a wide set
Of unknown future problems.

We are the un-kludged.

THE PROTON— LIGHTNING IN A BOTTLE

A proton is made of three quarks, true,
But quarks are so small
That they only make up
Two percent or so
Of the proton's total mass.

The quarks travel about
At near-light speed,
But they are imprisoned
In flickering clouds
Of other particles such
Other quarks and anti-quarks that
Materialize briefly and then vanish,
Plus, above all, gluons,
Which are massless and evanescent
But bind the quarks together
And carry most of the proton's energy.

Such, it would be accurate to say
That protons are made of gluons
Rather than quarks.

Experiments have proved
Wilczek's descriptions correct.

Every proton now around
Congealed from
A quark-gluon plasma soup
That existed for a microsecond
After the Big Bang.

In QED,
Which is about interactions
Of electromagnetic fields,
A screen made of photons
And short-lived electron-positron pairs
Partially cancels the electric fields,
Although, on the inside,
The fields get much stronger;

However, in QCD,
The opposite happens,
Called antiscreening—
The strong force between the quarks
Gets weaker as the quarks
Get closer together.

So the quarks seem to be
"Asymptotically free",
Moving about as if there
Were no force between them at all

But this freedom is an illusion,
For, as the distance between them increases,
So does the strong force.

It is that the proton has parts,
But it cannot be taken apart!

The proton is mostly glue,
A kind of a super glue.

One number remains constant in protons
As their virtual quarks and anti-quarks
Come and go in their uncountable swarms:

Three.

There are always three more quarks
Than anti-quarks.

Why?

Why is the universe made of matter
And not anti-matter?

A slight imbalance between
Quarks and anti-quarks
Already existed in the
Primordial quark-gluon plasma.

Why?

EMPATHY EXPLAINED, AND MORE

Why do we not eat all the scarce food,
Feeling the other's hunger?

Why are moods contagious?

Why do we feel the wish to dance
When we see another doing so?

Why do I feel your pain
When you cut your finger?

Why might I get an itch
When you scratch yours?

How are you "out there"
Felt "in here"?

How do your acts become mine,
And my acts become yours?

Mirror neurons!

Using very thin electrodes,
Experimenters measured
The activity of a single neuron
Of the premotor cortex
In a monkey grasping a peanut.

Amazingly, a bit later,
An experimenter
Grasped a peanut—
And the same
Monkey neuron activated
Merely by watching.

This was the beginning
Of explaining vicarious feelings—
The mirroring of others' behaviors.

It seems that free will
Is not so impregnable;
Each time I witness your movements,
You permeate my stronghold.

It extends to sounds, sensations,
And emotions, as well,
And so we can feel
All of those inside of us,
As if we were in another's shoes.

These brain circuits blur the bright line
Between your experiences and mine.

Without these mirror neurons
There could have been no learning;
But, it goes beyond that
And onto intuitive altruism.

In many places in the world,
People tend to share the wealth.

Of course, sometimes,
Our desire for benefits
Might outweigh our empathy.

In the military,
The General is at a distance
That separates him or her
From the suffering
That their armies cause.

The same for weapons
That kill at a distance—
Empathy can then be bypassed
In the service of efficiency.

Otherwise, it is
"Do unto others
As you would have them
Do unto you".

Each time we see an action,
Our mirror neurons mimic
And transform this sight
Into the motor commands
Necessary to replicate the action;
However, a neural gate
Blocks the immediate output
Of our motor areas.

Behind this gate,
We can covertly share
The actions of people around us.

We feel them,
And they thus become
A part of our extended self.

The brain is ethical by design.

It was advantageous
To know another's needs.

THE GALACTIC FRAMEWORK

From my perspective here
Out on the Orion spiral arm
Of the Milky Way Galaxy,
I can see a lot of it,
Although necessarily edge on.
But I don't see that it
Is doing much for us, as yet.

Couldn't we have gotten along without it?

Well, no, not if it produced our sun.

NO CREATION

It was impossible for there
Not to be "something",
Or the capability thereof
(Quantum tunneling),
Which is still "something".

So, then there would be no Creator,
Nor none even necessary.

This is because Nothing
Could not produce "something",
Nor anything,
As Nothing has no properties,
Capabilities, etc.;
So, "something" had to be, plus it is,
Plus, a lack of anything (Nothing) is not.

No Divine plan has been seen or is required.

Consciousness is obviously tied to the brain:
Anything talking to the brain
Would have to speak the same language;
Thus, no immaterial stuff of a "soul".

OBEs and NDEs are totally
Explainable by science.

There is no proof of past lives whatsoever.

Morality was/is provided
By humanity / natural selection.
The afterlife is but a wish.
Being "special" is not the ultimate freedom;
No strings attached is the ultimate freedom to be.

The wish is never so humble,
But ever hopeful and even greedy for reward.

"SMALL NEUTRAL ONES"

Their mass adds up to more than
We can see in the night sky,
Even through our telescopes,
More than all the atoms.

They mostly go in a straight line
Towards the edge of the universe,
Straight through any planets,
Stars, mountains, and atoms.

40 billion just went through your left nostril.

They are virtually unstoppable.

When one is detected,
It is called an event.

They oscillate,
Even changing to a different
Flavor and mass,
Making them even more undetectable

They live and die
But by the weak force—
Born by decay
And dying when
They become something else—
Being immune to the
Strong nuclear force
And the electromagnetic force.

Wolfgang Pauli made them up,
For bookkeeping purposes,
When a neutron decayed
Into a proton and an electron.

No one doubts their existence now.

They are all that can get through
Fifty feet of iron.

What are they?

(Neutrinos)

THE ORIGIN OF CURSING

It was said that God damned
A lot of people and things,
And so I guess people,
Being what they are,
Imitated "God',
The vengeful version
Who even broke
His own commandments,
Making Him not really
A good role model at all.

But cursing and damning caught on,
And so we have it that
Even God's last name [Dammit],
Ha-ha, was taken in vain.

Some would-be-but-not-quite swearers
Would also exclaim "Jesus H. Christ"
When annoyed,
Or "Jesus, Mary, Joseph",
As my aunt used to say.

Worse words then came about
As analogies to a person
Being some unsavory part of the body
As a way to get in a great insult.

For me, and some,
A new four letter word is W-O-R-K.

Shitake [a mushroom].

THE UNIVERSE IS TRYING TO TELL US SOMETHING

So, we will have to interpret it,
Seeking the final story.

While well-formed,
No one thought
That the Standard Model
Was the final word.

For one thing,
It doesn't include gravity,
And so it is incomplete,
As well as having
Varied symmetries,
Some working
And some broken.

Vacuum Energy

Vacuum energy is
The energy density
Of "empty" space,
And, by Einstein,
Is inherent in
The fabric of space itself,
Imparting a persistent impulse
To the expansion of the universe;
Whereas, ordinary matter
And radiation disperse.

Yet, the estimation
Of this cosmological constant
Is off by 10**120.

What is the universe telling us?

Well, for one thing,
If the constant really was larger,
We wouldn't be here.

Do we live in an unusual part
Of the universe?

Well, who knows, but,
What about the Big Bang?

The History of the Universe

We are talking here about
The Big Bang model,
Which is not about the Big Bang itself,
But about the afterwards.

While the Big Bang is
But a placeholder for our ignorance,
The afterwards Big Bang model
Is on firm empirical ground.

There are still some questions:

Why did the early universe
Look like it did—
Hotter, denser, and smoother?

And in the future the universe
Will have gone from the primordial soup
To but a thin gruel of particles
Growing even colder
And more distant from one another.

Thus, the past was
Very different from the future.

The arrow of time moves forward.

Higher entropy configurations
Are more natural than lower ones,
For there are a lot more
Of those configurations to be found.

However, if entropy had always
Been very high. or maximal,
There would have been no life.

We have a medium entropy.

Why did matter collect into galaxies rather
Than spread uniformly and thinly throughout space?

The Prehistory of the Universe

What! Huh, how can this be here!

Well, we will try to peer beyond the veil.

Imagine a previous, dispersed universe;
However, even empty space
Is not perfectly quiescent,
For there is still vacuum energy—
The irreducible fluctuations
In the "vacuum".

Here, particles pop into
And out of existence,
And the fields occasionally
Arrange themselves
In statistically unlikely configurations.

If we wait long enough,
The right kind of configuration will arise,
Giving rise to an entirely new universe,
As it pinches off a small region
And creates a disconnected bubble
Of distorted space.

It had to be a small region because
It was easier to create a new bubble universe
In such a configuration than it is
To make a large, dilute universe from scratch.
This may be what the universe
Is trying to tell us.

HIGHER CONSCIOUSNESS

The three lower consciousnesses that are
Obsessed with the securing of objects,
With the chasing of sensations, and with
Power/control will never ever be enough.

If we don't accept the unacceptable,
Then we lower our level of consciousness
Our response will mirror their uptightness—
Which can spread the bad moods onto others.

Conscious Awareness, which can but witness,
Is a safe haven from which to observe
The drama of our lives playing in our minds,
Granting us a sobering distance from it.

From a safe subjective place that's free of fear,
Our soul, our conscious awareness, can witness
The strange thoughts and emotions that surface
On the mind, sent by the subconscious brain.

Putting ourselves in the place of others
When hurtful things are done to us,
Expands our consciousness, compassion, and love
Since we can come to know why they did it.

When we converse with ourselves, it is our
Higher Consciousness—our Conscious Awareness
Or I, that questions our lower consciousness
Impulses toward securing, sensation, and power.

Seeing the big picture of life and its stages
And connections lets one not get annoyed, say,
At being cut off in traffic, for s/he
May be old, learning, lost, growing, or angry.

Putting the needs of others ahead of
Our own produces the byproduct of
Happiness and reduces stress, for we
No longer have unrealistic expectations.

DARK ENERGY

It's the only "stuff"
That acts both on
The subatomic scales
And across the largest
Distances in the cosmos.

Its aliases are
The cosmological constant
And the vacuum energy.

GR tells us that space and time
Do not together create a fixed, empty stage
Where matter and energy can dance about,
But, rather, that space-time itself
Bends in reaction to matter.

By Einstein, mass can arise
Either by acceleration
Or gravitational expansion.

Mass and gravitational force
Are both manifestations
Of the bending of space.

All states of motion are relative,
Including rest,
But what is absolute is
That space-time can be bent
In the presence of matter or energy.

Gravity operates without
The need for a background space;
Rather, it defines background space.

In contrast, the electromagnetic force
Requires space in which to extend.

The gravitational field
Is itself the background space-time.

At large distance scales,
The e/m force and
The weak and the strong forces
Are irrelevant compared
With the gravitational force.

Dark Energy, an omnipresent,
Evenly distributed, repellant substance—
Acts on gravity to make
The very fabric of space accelerate.

It has negative pressure
And positive energy.

Fields carry the four
Fundamental forces of nature.

Fields can be particles,
But particles return to the parent field.

Since the vacuum state
Is the lowest-energy state possible,
A particle cannot spontaneously
Be created unless its antiparticle
Is also created.

This bubbling of matter and antimatter
Generates an electric force
Called the Casmir effect.

The properties of the vacuum energy
Are none other than
The quantum manifestations
Of Dark Energy,
Which acts on gravity
To create more and more
Space at an ever increasing rate

Or, is gravity insensitive?

Are particles indivisible?

If we think so,
Then we maintain
The intrinsic divide
Between matter and space-time,
So, we might try to see
The vacuum as relative
And not as an absolute entity.

Perhaps what are thought
To be fundamental particles
Are the equivalent
Of a quantum emergence
Of the fabric of space itself,
Being just as relative as motion.

Dark energy would then be
An artifact of the emergence
Of space-time
Over larger distances.

OF QUIRKS AND QUARKS

Whatever physical laws lurked and larked,
Back then, parked, of the deep and dark,
Led to a very small excess of quarks
And leptons over their counterparts—
Of the order of 1 in 30 million sparks,
Resulting in the predominance of matter farts
In the present universe over antimatter arts.

From these marks we mammals harked,
With mind, soul, spirit, and hearts,
Although some other mammals barked
And some even quacked and quarked,
Floating in our wonderful Earthly ark
Upon life's ocean of dangerous sharks.

EDEN'S LONG LINE

Methuselah lived for 969 years.
He died on the 11th of Cheshvan
Of the year 1656
(Anno Mundi, after Creation),
7 days before the beginning
Of the Great Flood.

According to Rashi on Gen. 7:4,
The Holy One delayed the Flood
Specially because of the 7 days of mourning
For the righteous Methuselah in his honour.

The recent find of Austi 2:5 scroll tells us
That Adam was yet alive on that day,
He, too, boarding the Ark of Noah.

In fact, I ran into him just the other day,
Looking innumerable years old,
But aging quite gracefully.

Eve was at his side,
Yet gleaming with the ripeness
Obtained from Eden's apple.

They revealed the formula for true apple cider,
Which would result in an elixir not vinegar.

Eden's sinful Apple, the cause of it,
Made for harsh apple cider, but, when it
Was heated with sulfurous brimstone it
Soon turned smooth, the Hell taken out of it!

(Methuselah was son of Enoch,
And grandfather of Noah.)

EAST VS. WEST OUTLOOK EXPLAINED

Eastern ways have it
That interdependence is to be sought
As a frame for seeing the world
And one's place in it,
While in the West
An independent frame is stressed.

In essence, Eastern Asians
Are raised to believe
That we are all connected
And that the needs of the group
Outweigh the needs of the individual.

In contrast, people from Western Europe
And North America are taught
To prioritize their own goals,
Feelings, and achievements.

In Eastern cultures
"The nail that stands out
Gets hammered down",
While in Western cultures
"The squeaky wheel gets the grease."

These opposite ideas
Have endured for over a millennium.

Their brain organizations must differ
And so we will see that it does, and why.

There are three different forms
Of the serotonin transporter gene
(5-HTTLPR),
Based on the combination
Of two alleles, called short-short,
Long-short, and long-long.

While two-thirds of East Asians
Have short-short,
Only one-fifth of the West has it.

This gene in particular
Is related to socioemotional sensitivity.

Those with the short-short variant
Have been shown to be at risk for depression,
But only if they lacked social support,
While the short-long and the long-long
Variants remain unaffected
By the lack of social support.

The short-short neurochemistry
Would then predispose the East
To establish interdependence
As a cultural value,
One that makes everyone's
Well-being a priority.

As reflected in the East's
Way of life (kind of a religion),
Their culture solidified in the form
Of neo-Confusianism,
Which combined the Buddhist beliefs
That we see are all connected
And that selfish attachments are unhealthy.

Western culture emerged out of
The combination of Judeo-Christian theology,
Which posits a single God
Who holds individuals responsible
For their own eternal salvation,
Plus, from Greek civics,
Which emphasized personal agency
And free will.

Both of these Big Ideas
Migrated until they found
The population with the
Right neurochemistry
To make them stick,
Although some countries,
Like India, remain half-half.

So, while it's true that people
Must go with their brain orientation,
It helps to be informed
Of the whys, such that one
Will able to see both sides,
And to see that we are predisposed
To find some ideas more appealing.

Science ever finds things out
That can't be found by introspection alone.

THE FRIENDLY PLACE

No one here gets up in the morning
To look for what's wrong with the world,
But to see what's right about it.

All are abducted by good cheer
And not by alien spacecraft.

Even the Hare Khrisnas
Are happy and not pushy.

It is the warmth and the flowered air
That helps to lift the spirit.

Sometimes we have breakfast
At the shore with the birds;
Other times we sleep away the day
If it rains or it is chilly.

Cozy, cozier, and coziest;
Balmy breezes caress.

There are no deadlines, no clocks;
Just the lightness of being.

All nonsense is bypassed.
Money is piling up.
A brassy sea.
Sangria.

THE ALIENS ARE HERE
(THE VIRUSES)

You feel that you are not alone;
They are all around.

The near-invisible life forms
Of viruses swarm all around you.

The alien has integrated itself
Into the very fabric of life
That surrounds you.

There is no escape.
It has invaded and won!

Viruses are nanocreatures
That have penetrated
Forms of life on our planet
With startling efficiency.

They are merely genetic material
In a protein coat.

The dominant forms of life on our planet,
When measured in biomass and diversity,
Are microscopic.

There are 250 million virus particles
Infecting bacteria in every milliliter
Of unpolluted natural water ecosystems.

The existing equilibrium of our planet
Is dependent on the actions of the viral world.

20-40 percent of bacteria
In our marine systems
Are killed by viruses each day,
Which provides a tremendous source
Of organic matter, releasing amino acids,
Carbon, and nitrogen, recycling nutrients.

They also prevent any one
Bacterial species from dominating.

Because of their high mutation rates
And their ability to exchange
Genetic information with one another,
Viruses are tremendous generators
Of genetic variation.

The introduction of a retrovirus
Into our ape ancestors
Led to a new mammalian gene
That play an important role in our placenta.

The use of unadulterated vaccinia virus,
A variant of the cowpox virus,
Allowed humans to wipe smallpox,
Perhaps the worst scourge humanity
Has ever face, off the face of the Earth.

Like aliens, viruses are usually portrayed
As either perfectly benign or perfectly evil.

They are the ETs.

Viral Evolution

Is a subfield
Of evolutionary biology
That is specifically concerned
With the evolution of viruses.

Many viruses, in particular RNA viruses,
Have short generation times
And relatively high mutation rates
(On the order of one point mutation
Or more per genome
Per round of replication for RNA viruses).

This elevated mutation rate,
When combined with natural selection,

Allows viruses to quickly adapt
To changes in their host environment.

Viral evolution is an important aspect
Of the epidemiology of viral diseases
Such as influenza (influenza virus),
AIDS (HIV), and hepatitis (e.g. HCV).

It also causes problems
In the development
Of successful vaccines
And antiviral drugs,
As resistant mutations often appear
Within weeks or months
After the beginning of the treatment.

RNA viruses are also used
As a model system
To study evolution in the laboratory.

One of the main theoretical models
To study viral evolution
Is the quasispecies model,
As the viral quasispecies.

HOW OLD IS THE SUN?

In 1860, Lord Kelvin,
Who did some work on absolute zero,
Warmed up with an estimate
On the age of the sun at 30 million years,
Figuring it was made of molten rock;

Yet, there was one, a botanist,
Named Darwin, who knew better,
For a much longer time than that
Would have been needed for
Evolution via natural selection,
Which, ironically, Kelvin was opposed to.
(The sun is 4.6 billion years old)

PENROSE'S "MANY PLACES" EXPERIMENT

Sir Roger Penrose
Has though about something
For a very long time,
Ever since Paul Dirac told him in class about:
"...the superposition principle,
Whereby very tiny objects could be
In two places at the same time."

This blurry flux even allows
An "infinite" number
Of locations simultaneously.

Yes, quantum mechanics works perfectly;
But, what leads to the world
At ordinary scales?

What collapses the quantum wave function?

Penrose believes he has
Identified the secret
That keeps the quantum genie
Bottled up in the atomic world,
A secret that was
Right in front of us all along.

It is gravity.

The flaw in the Copenhagen interpretation
That collapse is due to "observation"
Is that it has no basis in theory.

Gravity is the only one
Of the fundamental forces
That physicists have been unable
To explain in quantum terms,
Einstein trying for 30 years,
This perhaps being a clue that
Physicists are on the wrong path.

How would gravity affect
An object small enough to exist
In the borderland between
The quantum world of atoms
And the human world of visible objects?

There should be such a place where
The quantum approaches the classical.

An object about the size
Of a spec of dust might
Provide the perfect test.

At this scale, an object is small enough
To be strongly affected
By the rules of quantum mechanics
But large enough to observe directly.

If there was a way to observe the spec
Without disturbing it, we would see
Quantum strangeness laid bare:
A macroscopic thing
Sitting in two places at once.

Quantum theory is incomplete
Because it ignores the effects of gravity.

Gravity is so weak on atomic
Or subatomic scales
That most physicists leave it out,
But tiny objects should,
By Einstein's theory,
Produce space-time warps, too.

If a dust spec is in two locations at once,
Each one should produce its own
Distortions in space-time,
Yielding two superposed
Gravitational fields;
Yet, it takes energy
To sustain these dual fields.

The higher the energy required
To sustain a system,
The less stable it is,
So, over time,
It tends to settle back
To its simplest, lowest, energy state,
That is, to just one object
Producing one gravitational field.

If Penrose is right,
Gravity yanks objects,
Perhaps above a certain size,
Back into a single location,
Without any need to invoke
Observers or parallel universes.

What is the degree of instability, though?

Electrons, atoms, and molecules
Are so small that their gravity,
And hence the energy,
Is negligible, and so they
Can persist that way "forever".

Very large objects,
On the other hand,
Create such significant
Gravitational fields
That the duplicate states
Vanish almost at once.

For a dust spec,
The process takes nearly a second,
Long enough that it may be measured.

Is there an experiment?

Instead of a spec of dust,
Penrose would use a tiny mirror,
Bouncing radiation off it
To see if it was in one
Or two locations at the same time.

If Penrose is right,
The mirror would maintain
A dual existence for no more than a second
Before gravity chained it to a single location.

He initially wanted to use
An x-ray laser mounted
On a platform in outer space.

It would shoot photons
Towards a tiny target mirror
Tens of thousands of miles away.

A half-reflective mirror,
Called a beam splitter,
Would separate each photon
Into two states
So that it would follow two paths
At the same time.

On one path,
The photon strikes the tiny mirror,
Moving it slightly;
On the other,
It is reflected away
From the target mirror,
So the mirror does not move.

In the prevailing quantum view,
Both events occur simultaneously:
The mirror moves
And remains in place
A the same time.

On its return path,
The duplicate photon
That struck the mirror
Hits the same mirror again,
Returning it to its initial position.
Since there is fundamentally
No way to tell which path
The photon took,

The two photons interfere with each other
And recombine into a single photon
That is always reflected along a path
Back toward the laser;

Thus, no x-ray photons
Can ever follow a path
That leads them to a detector
Which would be sitting off of
The first half-reflected mirror.

However, if as Penrose expects,
It forces the tiny mirror
To either remain at rest or move,
But not both,
Because gravity anchors
The tiny mirror to a single state;

Consequently, each photon
Will follow one path only.
So it cannot interfere with itself;
Half the time leading it to the detector.

Thus, the quantum duplicate
Of the mirror must have disappeared,
And so Penrose's view of reality
Would be the correct one.

Well, there is too much expense
In performing the experiment in outer space,
So, Dirk Bouwmeester
Has devised a way to bring
Penrose's experiment down to earth.

A visible light source is to be used
Instead of an x-ray laser,
Giving it the same kick
By reflecting the light photons
Back and forth between
Two mirrors a million times.

They are past buckyballs now,

Soccer ball-shaped carbon molecules,
in size, up to an organic molecule
Called azobenzene,
Although the tiny mirror
Would be a billion times bigger.

They are working on ways
To shield the experiment
And students are creating the mirrors.

Stay tuned for a few more years.

PHOTONIC INFORMATION

Referring to objects,
Photons come off the object.
Our brains use the angle
Of incidence of the photons,
How far they've come, their intensity,
And all that to re-present the object.

Other photons come from the sun
And the stars, objects as well.

FREE?

As for our choices, they are as wide
Or as restricted as our learning.
After that, well, we surely want our choices
To reflect what we have become.
The notion of "determined" doesn't sit well,
But if we consider its opposite—
"Undetermined"—then it does.

THE FUTURE PAST

Atoms of a type are identical,
But one radioactive atom
May decay well before another.

Yet, there is no definitive cause
For the different behaviors,
No way to predict the decay time
By looking at their histories.

What regulates the particles' behavior?

Not the past.

Where is the information,
If not in the past?

It can only be from the future.

Holy cripes!

Tollaksen and Aharonov
Designed experiments
In which the outcome was determined
By events occurring after
The experiment was done.

There were three steps:

(1) A "preselection" measurement carried out
On a group of particles,
(2) An intermediate "weak" measurement, and

(3) A subset upon which a final
"Postselection" measurement
Was carried out.

If information flowed from
The future to the past,
Then the effects measured
At the intermediate step

Would be linked to the
Final subset measurement.

A weak intermediate measurement was used,
So as to not disturb the quantum properties:
A motorized mirror whose movement
Could get amplified from the final measurement.

So, when a final measurement was made,
If it was, then it was seen that
The deflection angle of the mirror
Was amplified by more than a 100 times!

Somehow the later decision
To make a final measurement
Appeared to affect the outcome
Of the weak, intermediate measurements,
Even though they were made at an earlier time.

Is the future known?

Yes.

Did the future already occur?

Yes.

How much of it has occurred?

Well, I am not allowed to say,
For that is classified information.

HUMAN EQUATION = ?

Because we are free to be,
We are the equations without an answer.
We solve ourselves.

ON THE ORIGIN OF THE PHYSICS

Andreas Albrecht has killed time.

He was interested in a time
When the universe was compressed
Into a space the size of a grapefruit.

The equations of quantum mechanics
Became confusing when he tried
To tell the story in a linear way,
For all possible futures
Were described in terms
Of the probability
That any one would actually happen.

However, he realized that
There was no unambiguous way
To come up with a tick-tock version,
So he tried to manipulate the equations
To isolate time,
Even having to attempt to undo
The unification of space and time,
But still trying to jury-rig
Some kind of a clock.

But his models kept falling apart,
The equations always leading
To a different kind of universe.

He called the problem "clock ambiguity".

It seemed that the laws
Determining the history of the universe
Could never be specified beforehand
[Indeed, if causeless],
And that the fundamental physical laws
Were not fundamental.

So, then there would be
No timeless rules
That could be described by math,

No ideal realm of forms
That embody the laws of physics.

Smolin, too,
Feels that time as we know it
May not emerge from
Some deeper set of laws

Kauffman offers that
The laws of the universe
May have evolved over time,
Just as those of the biosphere
In which evolution in biology
Favors change from
The simple to the more complex.

Perhaps certain types of laws
Came to dominate
Because they are more successful
In producing a complex universe.

THERE ARE NO PSYCHIC POWERS

One good example of how we know
Is the million dollar challenge.

The James Randi Educational Foundation
Will award anyone who can demonstrate
A paranormal or supernatural power
Under controlled observing conditions
One million dollars.

So far no one has made it
Past the preliminary test.

You'd expect that if anyone
Actually had these powers
That they'd manage
To demonstrate something
In order to win the money.

THE SHORTEST POEM

Sunshine, fresh air,
Existence everywhere.

The above is far from being
The shortest possible poem.

Me,
Thee.

This is short, but probably
Not the shortest, as it has six letters.

Is this the shortest rhyming poem,
Especially about the TOE?

I,
Why?

How about

Me,
We.

Well, that ties the record
So now we need a three letter poem.

Aye,
I.

This again tied the record,
And it still makes sense,
Which of course is always a requirement.

How about?

Hi
I.

Is it not significant enough
Because it is only the start

Of a lower-higher self conversation?
Or do we know that from that?

Or, like the phrase repeated
at the Cheech and Chong show:

I
Hi.

or

By,
I.

Now we must reach
The two letter poem of two lines of rhyme...

I
"I".

Meaning, for some,
That I am the same as "I",
The soul or consciousness?

Or to question it the other way around?

I
"I"?

(But they both suffer from the flaw
Of the rhymes being the same word.)

W,
2.

Not in the IRS sense but, double you, too.

Then, finally,
There is the 0-length letter poem entitled

"The Zen Poem of Nothing"

(Hey, it's not there, but it really is,
Since nothing rhymes with nothing,
Plus the poem has a title,
Which makes it qualify as real.)

THE DRIFT

I drifted down the River of Eternity
Born and borne upon its meandering current,
Floating, bobbing, and wafting unto this life,
Now here wandering, puttering, and dawdling,
Often digressing, deviating, diverging,
Veering, and getting sidetracked.

All eventually piled up, banked up, heaping up,
Accumulating and gathering the bulk amassed,
Yet ever shifting, flowing, and blowing,
And relocating, until I caught the drift...

Of that gist, essence, and meaning,
The drift of the shift,
From the sense, the substance,
And its significance
As the mound that accumulated into me—
That thrust, tenor,
And implication imported into me,
Which spurred the intention
To direct the course of history.

COSMIC RAYS

Some cosmic rays are so energetic
That they must have been born
In cosmic accelerators
Fueled by cataclysms
Of staggering proportions.

They should not even exist.

Yet, they do,
As discovered 17 years ago.

One had 3x10**20 eV!

Yet, they do not travel well,
So they must come
From relatively near by.

They are also exceedingly rare,
So we spread counters
Over 40 square miles
To detect the secondary
Scintillation particles.

Auger researchers have linked
The ultrahigh cosmic rays
To active galactic nuclei
Close to home
That are thought to be
Powered by black holes.

THE FOREST OF ORIGINAL GROWTH

What would it be like to stumble across lands
That no one else had ever been to,
And how could you know that?

After reading Sir Conan Doyle's 'Lost World'
About dinosaurs on a sealed off plateau of a volcano,
I wondered if there were any more undiscovered places
That the paths least followed could lead me to.

So, while at the Earth Summit in Rio last month,
I forayed into the uncharted regions of Brazil,
Having chosen from a map the remotest area.

After various vaccinations and preparations,
I trucked my one-man helicopter
To the last way station,
Loaded the extra gas tanks onto it
And flew into the heart of darkness,
Eventually gliding down onto a grassy field
Just as the gas ran out.

From here I walked for tens and tens of miles,
Always taking the most difficult path
Whenever there was a choice,
For this would insure that I could end up
In some totally unvisited region
That was near impossible
Or hard-to-get-at in any way.

After hundreds of these
"Improbable" path choices
I suddenly came across acres
Of Lady's Slippers flowers.

These are very rare flowers
That usually appear in small bunches,
Growing only in conjunction with a rare fungus,
And, even, so, usually get picked—
But there were millions of them.

After taking one last really difficult choice of a path,
I discovered entire fields of other flowers
Long thought to be extinct.

Some were Eve's Blossoms,
Which not been seen for thousands of years,
Historically valued for their life extending elixir,
As well as the original, lost, strain of Pearly Everlasting,
The flower that never dies, and so I suspected
That I might be in virgin territory.

How would I know?
Well, for one, there were no paths left,
For even animals and their hunters
Had either long left or had never even been here.

Also, the flower colors were not like any
That I had ever seen before,
Not new colors, mind you, but, just, well,
Colors of different intensities and hues
That were not thought to exist in nature.
I saw true-blue roses, legendary no more.

I had chanced upon a land
Of strange rainbows of elfin-hued flowers:
Red Delphiniums, Black Tulips,
Orange Fuchsias, White Marigolds, Bronze grass,
Yellow Violets, and even Adam's Apple,
Now growing from the ground!

Was this the original forest—the Garden of Eden?
Was I the first to return to this legendary paradise?

And then I knew that it was indeed the Garden,
For there, right in front of me,
Was a field of thousands of undisturbed
Golden nuggets on the forest floor.
Surely no one had ever been here,
At least not for a long, long time.

I reached up and put the apple back on the Tree.

TO THE END(S) OF THE UNIVERSE

I took a road trip
Through the universe recently,
Smoking some pot
And playing the radio loud.

Holy-moly, there's nothing holy out there.
In fact, it's a very uncongenial place for life.
I'd much rather be in Australia

96% of it was useless
Dark energy and dark matter.
The rest was mostly rocks gases and dust.
Dangerous radiation zapped all over the place.
And it was fricken freezing!

Oh, what I would have given to be in Canada.

Whatever designed the universe
Certainly didn't have life in mind.
It even took evolution billions of years
To fine-tune us to the earth.

Then we nearly got wiped out
By huge disasters right and left,
Even once shrinking back down
To a population of around 2000.

I saw the graveyards of stars
And some stellar nurseries, too.
All kinds of energy swirled about—
When it wasn't exploding and wreaking havoc.

I stopped to eat at the restaurant
At the end of the universe,
On a moon,
But it had no atmosphere,
Plus all the food had been microwaved,
By the CMBR.

What a wasteland
Of a wilderness of wilds
Of a whole bunch of crap
That nearly goes on forever
In every direction.

This was as much of a place
Unsuited for life that there ever could be.

I'm back, thank my lucky stars,
Noting that, 14 billion years
After the initial chaos, here we are,
Having beaten the odds.

Well, someone had to!
We won the universal lottery jackpot.

Oh cripes,
Here comes a humongous asteroid!
Darn, all that luck for nothing.
Double '00' has come up.

It was only a matter of time.

NO GO

If "He" used cosmological processes
And evolution as means to His ends,
The the means He used are unbefitting
An all-powerful and all-loving God.

What we do know is that
There is ample imperfection
And misery in the 'design' of nature
To justify the conclusion that
The Creator is either malicious, incompetent,
Indifferent, or simply nonexistent.

FLUID DYNAMICS

Two hints of a wisp of a breeze,
Each the other way going,
Passed near each other,
At the equator, this time swirling
And continuing to spiral,
Ever more and more,
Eventually giving birth to the beast
That was Gustov,
The blob that ate Santo Domingo
And spit it out,
After raking Jamaica lengthwise,
End to end.

The whirlpool, a category 4 hurricane,
Drew energy from the water,
Then grew larger in mass,
Which then produced more energy—
A feedback system out of control.

It clipped Cuba's feet
And drowned the outer Keys,
Heading now for the warm waters
Of the gulf to refuel and perhaps growing
Into a humongous monster of 5 or 6.

Gulf oil platforms had been emptied,
And the roads of the coastal towns
Were all one-way out, both lanes,
No return possible and none desired.

The new New Orleans
Awaited the hand of fate,
Tempting doom
To chance its way once again.

So to, perhaps,
Our universe of whirling forms
Began as such,
As two oppositely flowing streams
Of fundamental energy-substance meeting,

And then whirling/twirling
Into spinning nucleons
That threw off photons
At their own spinning speed of light;

Then larger forms whirlpooled into stars,
And galaxies turning,
The voracious beasts of black holes
At the center devouring all those
Who entered the Gates of Hell—
All hope having been abandoned,
A one-way street to oblivion.

INANIMATED TO ANIMATED

Perhaps the knowledge of movement
Makes for animated life,
But when did this happen?

How does one draw a clear line
Between organization and not?

When, even, does the night turn to day?

The most interesting and potent things,
From the evolution of the universe to life,
Exist at the blurred boundary
Between order and chaos...

Life perhaps emerging in tide pools—
The shifting edge between land and sea.

It is all of the fuzzy realm
In which and where things
Have to be orderly enough
To take form,
But not so much frozen
That they cannot change.

MAGICAL HAPPENINGS

What secrets of life and death
Lay buried in the sands?

What inaccessible truths
Protect themselves by their own magic?

Old Rascal lit up a cigar,
And the stories unfolded
In the haze of this pipe dream...

"Do tell what else
Was in that Great Pyramid, Fredrick,"
The General suggested.

"There were 4000 year-old iron weapons
That did not rust,
Looking as new as the day they were forged.

"I held glass that bent without breaking.

"I drank from a vase
That poured water without end;
I filled an entire tub from it
And bathed away all my dirt and dust.

"A compass needle went around
And never stopped.

"I ate a cake but I still had it.

"I saw the starry skies
Through solid rock walls.

"I entered a room that had no door.

"There was light within the room
But no flame or openings.

"I looked into a grain of sand
And saw eternity."

Fredrick paused, recalling.

"Outside, I saw the Sphinx.
Its glance was fixed on something else.
It was the glance of a being
Who thinks in centuries and millenniums.
I did not exist and could not exist for it,
For it was the face of eternity."

No one spoke.

The General rose.
"Next, after an hour break,
During which you might go out
To see the scenery,
We may hear some about a long trek
From an escape from a Soviet prison
Through the mountains
And across some ink-black rivers."

Questor and Top Secret headed down one
Of the many paths of Niihau,
Its secrets ever shrouded in mist from above
And all around from the other islands;
But, here they were,
Within the Forbidden Paradise...

"Come back, friends,"
Said the General,
"To hear of the dark,
The light, and the never."

"We are here, being ever."

"There are books unwritten and never told."

"We can listen until we get old."

"By what muted shore of the dark river
Did its strand call me forth?"

"We're sure that we'll never hear worse."

"By what far edge of furrowed forest
Didst the Motherland seek my name?"

*"Oh, Dragon, through what hazy depth
Of Gloom hast thou tread and threadest?"*

"Gather thee round and you shall knowest."

A-VOID

Ain't nothing there,
No place, no space,
No influences, no e/m,
No where, no how, no what,
No who, no then, no when,

Or, there is no place empty of field.

Either way, the void doesn't exist.

Before the beginning of particles,
There was not the Void
But the capability for the particles
Emitted or congealing.

And the Void was nothing,
Which is true, for it is not there,
And so there was nothing being within it
Such as a point.

When nothing comes in to us
Or our instruments,
Then there is nothing out there.
If something was there,
Like an amount of space,
Then this is what what would come in.

I see a galaxy afar.
The space in between is expanding.
Avoid the void and write good checks.

REWARD

The reward model reinforces anything
That can be considered supportive
Of a preexisting belief, or belief system,
Including but not limited to
Sensations and data/information.

For some folks,
God is rewarding, theism is rewarding—
These ideas have been coded
In the reward system
Via exposure during development—
By parents or community, or for other reasons.

The belief, the conviction,
Is certain in these minds due
In part to the reward system
(For a more advanced discussion about certainty
See Robert Burton "On Being Certain:
Believing You Are Right Even When You Are Not").

What I am getting at here
Is that there is going to be
A degree of certainty involved
With interpretation as a result
Of the reward system—
And that interpretation is hopelessly biased.

This is independent of whether one
Actively seeks certain experiences
Due to the "interpretational" nature/bias
Of sensation or experience.

This accounts for why we would see
An imbalance of mystical claims
For those who are predisposed to such beliefs,
Be they buddhist spiritual cosmic unities
Or ideas of God.

EXISTENCE OVER ESSENCE

We are back at the Oahu mountain base.
A cat has adopted us.

I may take a vacation from
My holiday from retirement,
Leaving the tour of the Big Island
For another time,
By just lazing around;
But, then again,
I would only have all the more time
To read and post.

These are very lazy days now,
As we sit in the shade on lounge-chairs
About twenty feet from the edge of the cliff.

Fort Shafter lies below,
With the city and the ocean
Much farther out in the distance.

The lady usually paints while I read,
And the cat perches at the very edge,
Looking out over all creation.

We cooked a prime rib on a gas grill, somehow,
Each of which we obtained from the PX.

So, there is food, lots of sleep, and love,
As well as ProfPat's spirit, earth, and moon.

The days and night are about 12 hours each
And the seasons never change.

The scents are on the breeze
And the life is in the living.

The absolute essence is of no real concern,
It being the uncaused tiny and simple of so long ago,
Something not very amazing,
As it is just some miniscule movements.

It is enough to be informed by science
Of that which has occurred in our universe
And all around us, up to now.

Each person has to make an ongoing life,
And so that's what's first and foremost,
And way beyond the pondering
Of the original essence.

To speak of life in its positive aspects
Is ever of real and immediate use,
As LabelWench often posts;
Negatives, politics, sufferings, and all those
May still instruct, as well,
But, I leave that to others.

The transcendental moments ever come,
Those filling up the scene within,
And, from without,
All the adventures of life's living.

The afternoon sun shines,
A thousand nuclear bombs worth
Going off in it every second,
It still having 5 billion years left.
We are a safe distance away.

Dinosaurs still fly, as birds,
And the bacterial kings of forever
Are still with us.

Sometimes we imagine
The graceful forms of the *australopithecines*—
Those who are yet in your heart and ours.

"VOID" NOT VOID OF CAPABILITY

The idea of the Universe
Being the result of
A Quantum fluctuation
Was first proposed by E.P. Tryon.

At first his proposal was greeted as a joke.
Today this idea dominants Quantum Cosmology.

If you look at what
A Quantum fluctuation
Of the vacuum is, it has
A particular character.

For an example a fluctuation
Of the electromagnetic field
Is modeled at the creation-annihilation
Of an electron and a positron,
(Accompanied by a photon
Connecting the creation
And annihilation vertex).

Because this is a creation
Of positive energy and negative particle
This process involves no violation
Of energy conservation
At the micro level.

The energy and momentum
Sum to zero at each vertex.

WORLD GOVERNMENT

The global Ninja Empire will decide
Whether to neutralize nuclear launches,
And perhaps even the whole country of origin.

A complete breakdown of all electronics
Will be caused by
A giant electromagnetic pulse (EMP).

The implications of such an event
Will be enormous.
Phones will not work.
There will be no way to find out
Via the internet what happened.
All electricity will stop, etc.

It became clear only in 2002
That during the Cuban Missile Crisis,
The USS Beale had depth-charged
An unidentified submarine
Which was in fact Soviet
And armed with nuclear weapons,
And whose commanders argued over
Whether to retaliate with a nuclear torpedo.

On January 25 1995,
Russian President Boris Yeltsin
Came within minutes of initiating
A full nuclear strike on the United States
Because of an unidentified
Norwegian scientific rocket.

BEING FLING

Our roseate hearts are cleansed by the dew,
And lucky are we if the day finds us new;
As every blossom on the bush blows full,
We hail the wonders that morning bestrew.

Spring grows a clutch of blossoms to propose;
A zephyr blows nature's page to disclose:
Spring, departing, caresses the summer
And from this one kiss blooms the summer rose.

Spring's last breath awakens him, he's living:
The life-force passing to summer from spring.
His clover spreads, vines grow strong, roses cling—
All from the kiss of which spring dies giving.

The rose has thorns to keep the beasts away;
As such they preserve the fragrant bouquet.
Her petals unfold, meeting the light of day;
The queen of flowers melts my heart away.

Life's hardships can be softened by beauty;
Its weaknesses can be strengthened by truth.
As roses blossom like realizations,
Beauty itself blooms from the well of truth.

Soft breezes blow, caressing us two
As we kiss the roses and drink the dew.
Reason and passion soon merge into one,
As truth and beauty make their rendezvous.

The rose is the flower that the bee cruises,
Meeting there the butterfly that love chooses;
We unfold the petals of the blossom,
Then drink the nectar of love's sweet juices.

Her scent is ripe and her name means nectar.
Exotically blossoming I found her,
And buzzed my way into her flower,
For I was the bee and she my partner.

The Rose was pure white when it first was born,
Until Eve kissed it with her ruby lips
Or 'came it red when Venus fell on a thorn,
Rushing to the aid of struck Adonis?

Or did the Rose sprout forth, all fully blown,
From the heart of a Goddess, do you think?
Or was it out of Cupid's nectar grown,
When he poured to Earth that Heavenly drink?

Or when the nightingale, with hope forlorn,
Overpowered by the Rose's perfume,
Impaled himself in love upon her thorn,
Then revived in the beauty of her bloom?

With the Rose the Earth is rich forever—
It's born from spring's dying kiss to summer;
It wears all the gems that the dew has wreathed,
Blooming wherever summer's breath has breathed.

MIND-BRAIN

A 100 billion neurons exist in the brain,
Each connecting to about 1000 others,
Granting a lot of processing power.

We even deal with the animate (other minds)
Differently than the inanimate,
Leading some to think that there
Is even a dualism between
Our own minds and bodies,
But this has never been shown.

Even identical twins develop
Different memories and directions,
Showing that everyone is quite unique
In the history of the universe.

RIGHT AND WRONG

...as all people are truly aware
Of right and wrong action.

Yes, I'd say that most are,
And that they can still do
A cost-benefit analysis,
Such as *I can continue to steal*
But I will probably eventually go to jail for it,
But I'll still do it since it's easy money.

As this is what they've become,
They are true to their will,
As they must be,
Albeit it a fixed and limited will
That stagnated or grooved
Into some bad behavior pattern.

I'd say that wider learning
Always provides more informed choices,
But at the end of the day
We'll always want our will
To reflect who we are,
As its opposite would be "undetermined",
Which sits even less well
Than having no free will,
For it would have us doing "random"
And inconsistent things.

Some may be limited
In their learning and reasoning abilities,
Forever doomed to bounce off
The wall of life again and again
With little or no progress.
The more they get enabled,
Then even more
They wouldn't see a need to change.

So, can we will that which does the willing?
Well, it seems to be subconscious,
But at least it depends on

Our very own memories, learnings,
Personality, and associations.

Simpleton "wills" may even issue forth,
But the global brain at least might say,
"Nope, what the heck were you thinking!"

This kind of talking to one's self
Is the higher consciousness
Speaking to the lower, expressing "free won't",
But this, too, is also of what we've become.

We wouldn't want to suddenly
Turn into someone else,
But we may do so over the years
If we are receptive to learning.
If not, well, then, bad luck, I guess,
And life will fill with turmoil.

So, with learning,
We can will something tomorrow
That we might not will today.
Still not truly "free" but better than air-headed,
Plus, the results will be more creative.

OF WHAT IS A THOUGHT MADE?

The precursor of thought
Is you—your aspirations,
Your memories, your associations,
Your learnings, your methods,
Your personality, your direction,
All that you have become,
And your present state of being.

One must take care in what one wishes for
If the consequences have not been considered.

Unspoken thoughts
Become a part of the repertoire
That contributes to future,
The very purpose of thought.

?

"God" is but a larger question,
And certainly not an answer,
For then one must then explain it.

Thus, one has put a greater distance
Between us and the knowledge
Of the beginning of secondary particles.

This does not close the gap, but opens it—
As the human mind arbitrarily declares
The dogma of "stop" for its own satisfaction.

This only perpetuates ignorance
To future generations.

The human-written Bible
Is but human-written.

Pronouncements and declarations
Are nothing more
Than pronouncements and declarations.

If indeed logic and data
Could tell us that God exists,
Then there would be no need
For 'faith' and no unbelievers.

ONWARD

We have found the TAO
That gives rise to the now,
And why superpositions collapse,
Due to gravity, perhaps;

So, who would like to know
How the prime numbers grow?

It's the greatest secret ever kept
But I know its concept.

THE NEW TRANSCENDENCE

The fundamental proposition of materialism
Is that matter is the only reality.
Therefore consciousness is nothing
But brain activity.

There is the new vision of transcendence
Coming out of neuroscience
It's long been observed
That intelligent organisms
Require love to develop
Or even just to survive.

Not coincidentally,
We can readily identify brain functions
That allow and require us
To be deeply relational with others.

There are also aspects of the brain
That can be shown to equip us
To experience elevated moments
When we transcend boundaries of self.

What happens as the implications
Of all this research starts suggesting
That particular religions
Are just cultural artifacts
Built on top of
Universal human physical traits?

SEEING AN APPLE

(Brian Silston series)

The problem is we can't see apples
Absent a retinal change

Technically no, we can't see them,
But we can bring them into awareness via intention
Or due to subconscious processing that does it for us.

There are thus two ways
To bring an object into awareness,
One via intentionality,
And the other by subconscious processing,
Or unintentionally.

I don't see why there is so much ado
About one and not the other.

In any event we can see the apple
Represented in our mind
And whether the apple that we literally see
Or the apple our mind has conceptualized
Is experienced is a question up for grabs.

We're not sure, the answer may be both—
The pattern which we actually see
Will tell us some specifics about shape, size, etc.,
But this appears to get mixed
With our conception and representations
Of apples in general
And so that which rises to awareness
Has undergone some additional washing.

This fact explains in some types
Of illusions and hallucinations—
Can we be sure of what we saw?
Sometimes the answer is no—
Awareness can be duped.

Due to the possibility of a deterministic picture
Of human action and decision making,

Intentionality becomes a slave
To experience and interaction
Up to the moment
Where intentionality appears to be occurring.

This would put intentionality under the realm or rule
Of being the product of experience up to that point,
And would make intentionality
Not quite what it seems to be.

So while we might believe
We are making free willed intentional decisions,
These cannot be any other way.

The data that we have indicates
That the combination of many factors
Could explain decision making.

Metanalysis has its flaws but it is still based on data—

One of the criticisms is that
Any number of the studies being combined
May have different controls or standards
Which could lead to dubious conclusions.
But we should stick with what the data tells us.

The placebo effect shows us
That belief about something can change
How our brain reacts physically,
Releasing endorphins to lessen pain,
But I don't think this is intentional
From the agent standpoint—
We can't decide to tell our brains
To inhibit only glutaminergic neurons
Or allow acetylcholinase processing.

Likewise we can't will ourselves
To escape from depression on call.

Variations exist in the population
As to the extent one can
Intentionally control anything.

This suggests a physical aspect to intentionality
That I don't think we can afford to ignore.

It also suggests that intentionality has subcategories
And we must be clear which we are referring to.
Nonetheless there is a limit to how this thing
We are calling intentionality works.

This supports an evolutionary explanation
To the this phenomena.

It is not yet well understood, and sure,
We must be observant of the subjective conditions
To keep context, but I don't think we can rest assured
That the answer is a simple conclusion
That awareness is not physical at this time.

The subject-object example becomes confounded
Due to memory and mental models—
The brain recalls and processes objects
It is familiar with differently than novel objects.

We are more likely to see what is "really" there
When we see something for the first time
Because typically not much contextual information
Processing biases our sensory processing,
Which is what happens when
We view a familiar object or scenery.

This becomes further confounding
When the brain does attempt to "fill in gaps"
And match the perception to something
It is familiar with, which essentially fools us.

Thus that which is perceived in awareness
Is not a pure perception—
It's been corrupted by the contextual processing
Prior to arriving there.

Significant corruption in this system
Can cause us not to be able to tell
What is real and what is not real,

So awareness doesn't necessarily help us
Understand the world or get to any truths.

The auditory analog is interesting—
Activation is more prominent in the right hemisphere
For someone listening to, say,
Tchaikovsky for the first time,
As it is processed as one entity
Rather than broken down into constituent parts.
Those who have listened to Tchaikovsky extensively
Are far more likely to reveal activation
In the left hemisphere,
As the melody is abstracted into more symbols.

That aside, I maintain that experience is biased.
We recognize this as true and do our best
To get around it by inventing things
Like the scientific method to make those
Reviewing our analysis comfortable
That we are invoking the minimum bias possible.

In fact, if we were always biased and in error,
We could never know we are sometimes.

And many times we are
Until someone points out our errors to us
Or we come across something
That brings our errors to our attention.

The mind can cause physical changes in the brain,
As the mind is of the brain.

Neurons that fire together, wire together;
Sometimes this is controlled intentionally,
Sometimes we cannot control it.
Meditation is one way to induce these changes.

I would point out that intentionality
Does not seem to play as important a role
As we would like to think,
And I must assume we would
All like to think that it does.

Primed

With respect to intentionality,
It is reasonable to assume it is primed,
Either by the current environment
Or by the product of experiences
Leading up to that point in time.

We are ignorant of all the states
Leading up to this point —
But if we could know the initial conditions
And track the system up to some arbitrary point,
We could determine all of the actions
Taken by some subject.

The fact that seemingly endless options
Are available to us I think is illusory,
Or stated differently,
While there may be endless options
That we can contemplate,
It is not true that we will choose any one of them.

Most do not understand this. One person stated,
"Well I could choose to simply do the opposite
Of what my personality would suggest,
Thus confounding the theory".
I retorted,
You forget that your decision to do so
Includes our current conversation
As part of your experience,
And often present experience
Is a stronger predictor of choice
(this explains impulse buying,
Or home shopping or infomercial purchases).

The same with raising an arm
During a "free will" discussion.
I might even do the same,
But it's still my reaction based
On the discussion.

Intentionality

It is clear that what we are calling *intentionality*
Can cause physical changes in brain structure,
But there is nothing special about intentionality
With respect to changing structure—
For, this happens when
We think about something often,
Whether intentional or not.

It turns out that neuron nuclei release factors
Which reach oligodendrocytes and schwann cells
(These glial cells can detect action potentials
In some manner despite not having the capacity
To receive or transmit voltage signals)
And cause "greater" degrees of myelination
For certain circuits or loops.
Likewise the lack of use,
What we can deem to be lack of motivation
Or any intentionality or any thoughts
Can result in atrophy of brain regions.

Intentionality is possible due to our awareness
Of both internal and external states,
Or internal states as the case may be.

We can change the degree
Of perceived intentionality in a subject
By administering drugs, DBS, and other methods,
From inducing lack of motivation to creating it.

Thus, intentionality is in some way
Tied to motivation—
Which is further tied to other states,
Processes and so forth.

This suggests that awareness is critically
And completely dependent on the bottom up—
Cells with genes which link to systems
With processes and so forth.

We don't have any evidence telling us
That it exists in the absence of these things,
So it is very reasonable to conclude dependence,
Over invisible schemes merely pronounced.

It's difficult to then argue that despite this
It has its own phenomenal explanation
Separate and apart from that
Which appears to give rise to it.

Subjective Awareness

The subjective awareness we gain
Can be viewed as a positive symptom
(for lack of a better word).

Let's think of this positive symptom
As compared to positive symptoms
In schizophrenics,
Where those positive symptoms simply mean
Additional symptoms, or "phenotypes",
If you will, on top of cognitive deficit.

Subjective awareness does not occur,
As far as we can tell,
Until humans are
At a certain point in development.

This is why we don't remember being an infant,
And if we think we can
We are almost certainly mistaken.

Awareness at this point can be viewed
As below the level of consciousness,
Yet the brain remembers the sensory experiences
Which are encoded in implicit systems,
But there is no episodic or declarative memory.

The human then builds the mental model
Of the world over its brain architecture
And at some critical point the human becomes

Subjectively aware whereby
They can recall episodic memories
And make associations with them,
Including feelings,
Or being aware they are aware etc.

(This is opposed to Nietzsche's view
That subjective awareness,
Or controlled consciousness,
Arose out of being forced to turn inwards
Feelings of aggression or instincts
Due to the suppression
Of a more dominant group.)

I suggest this awareness is a positive symptom
That shows up at this critical point.
This leads me to believe that the emergence
Of subjective awareness is at
The intersection of these systems
At a critical point in development,
Perhaps the point where glial cells
Have largely finished pruning connections
To a certain critical degree
And some level of myelination is attained,
So that the connections among these systems
Can furnish meaningful subjective representations
Both within and towards the external world.

The subjective aspect is also proposed
To be completely dependent on the history
Of development experience added to
Any genetic modifications or influences
That may exacerbate what we call feelings
Or increased/decreased subjective awareness.

It is therefore biased in a way
That is undetectable to the subject,
Making it truly subjective
(There is no objective subjectivity
Unless you abstract away,
Which one cannot do and still explain effectively).

This explains seemingly unexplainable fears,
Like for instance fear of dogs
Which ends up being the result
Of a dangerous episode
During the developmental phase
Where episodic memory could not be formed
But implicit memory automatically reacts.

It is also the case where subjective awareness
Can be removed in clinical settings—
Affecting perhaps the communication
Between the systems described above,
Thus removing awareness.

This is a suggested plausible mechanism
Leading to subjective awareness
That can be tested when we acquire
The knowledge necessary
To pinpoint some of the critical points.

I certainly suppose there will be
Other suggestions as time goes on,
And those will be tested,
But I think this is
A plausible explanation nonetheless.

I suspect if we nail down the critical points
And necessary conditions of the beginnings
Of declarative memory we'll be closer
To an answer to this question.

Analyzing the subjective data
Is a problem in itself.
I don't think we (in general) could come to
Some agreement on how
It should be interpreted
And what it means,
So we are left in an eternal gridlock.

I don't see how we can
Take mystical experiences
At face value and be able to find

Some objective conclusion
That involves anything other than
And in no particular order—
The human brain is known
To produce subjective experience;

Every normally functioning human has a brain
And subjective awareness that operates
Across the same structure, cells, processes etc.;

Being susceptible to the same diseases,
Having the same sensations,
Having mystical experiences being
The result of this commonality
Of structure, cells, etc.

Instead we can pursue the question
Of subjective awareness as a phenomena.

We need to let neuroscience mature,
Let models be selected by time and effectiveness,
Let the community evolve and branch out,
Acknowledge our progress
And the inability to explain certain phenomena.

"The topic of awareness
Will be dealt with exhaustively?"

That is not currently the case—
We are still in adolescence as a field.

Only in the last 15 years or so
Has the field been able
To seriously begin advancing
To new levels thanks
To advances in technology.

I am embracing the physical solution here
Because emergent or otherwise,
Awareness occurs over
Our physical architecture
Via physical processes

(including chemical signaling)
And in no other way (yet).
No processes, no awareness.

Because our ability to actively
Measure mystical experiences
Is nonexistent I'd say
The data is completely subjective.

In fact that data *is* completely subjective.
It's about as good as putting someone on a couch
And doing a psychiatry evaluation.

All we have to go on in X felt this, X said that.
This is just not very convincing to me.

Consistency of various components
Doesn't tell me there is something more to it
Than some type of activity.

The fact that there is some kind of consistency
In reports suggests some type of process occurs
In some % of the population,
Just like OBE's or NDE's,
Or many other phenomena
That occur across cultures.

This is possible because
We have the same architecture
And process the same way.

The data here really isn't data —
It's a group of subjective descriptions
That have not been measured.
I don't think this tells us much at all.

"To have an experience is precisely
To be aware of what is happening to me."

Of course what we by this
Is having a unique experience
That has been encoded into memory

And which is retrievable without difficulty,
Which requires the hippocampus.

It is the case in persons without a hippocampus
(like the famous HM)
That according this this definition
No experience is possible because
New memories cannot be formed.

Now we may observe such a person
Having such an experience,
And even discuss the event in real time
But the experience cannot
Be recalled within minutes,
Or hours into the future.
I think this fact has some real implications
About the claim of nonphysical reality
Of mystical sensations.

"There is also an intentional subsystem
Responsible for awareness
And it is methodologically inaccessible
To a strictly third-person approach."

I suggest that theory of mind does a reasonable job
At providing a third person first person access
To the subjective awareness of another.

If we will insist that TOM is an approximation,
I would agree, but a rather good approximation.
Accurately determining intentions
Is an activity for which we have become proficient.

This would be impossible
Without unique or specific knowledge
About the subject for which
We are trying to decipher intention,
Or said another way, a representation about
The inner subjective state of another.

This is derived by a combination
Of our own experience

(both explicit and implicit inputs),
Interaction and observation of the subject
(like best-curve fitting),
The current circumstance
(immediate environmental factors
That may influence behavior),
Among other things.

That said, subjective self-awareness
Is not well understood currently,
But I don't think this means that
There must be something "unnaturalistic"
About what is claimed to be introvertive experience.

We do know that subjective experience
Is influenced by implicit processes
Which in turn are based upon our mental models
And life experience up to the present.

"The idea of reward-encoding
Will not explain how it can be sought
By those who have not experienced it before,
Or why it should come unsought."

I think this oversimplifies the reward model.
Let's assume that X is an atheist.
Each time X comes across information
That can be interpreted as supporting her assertion,

X will likely interpret that information
As supporting of her assertion.
The reward model reinforces anything
That can be considered supportive
Of a preexisting belief,
Or belief system,
Including but not limited to
Sensations and data/information.

...

Mysticism

The brain must code the memory—
Therefore must conceptualize
According to its algorithm
And if it doesn't we should
Be able to detect this
Via the new neural network associated
With the mystical experience.

The mystics may very well be
Tapping into emotional memory,
That which is ineffable,
And not able to provide
The factual aspects of the memory.

These 2 aspects can be separated
And this is done in patients with PTSD.
Aside from this the mystics would have to
Overcome one of the most important
Discoveries about memory —
Each time a memory is recalled
It is affected and re-stored differently
Than it was before.

We are getting a derivative of the last time
We recalled the memory each time we recall it.

By the nth time we
Have recounted a memory,
We are n+ degrees away
From the actual experience.
This is why treatments of recall
Coupled with the inhibition
Of areas like the Amygdala
In PTSD patients are effective.

I suspect the amygdala
Is quite active in those
Undergoing and recalling
Mystical experiences.
We simply cannot be certain

That these mystic experiences
Affirmatively prove or disprove
The existence of a specific deity.
This is where faith comes in, as I see it.

My inclinations tell me there is an explanation
For these experiences that involve mechanisms
We may be able to harness in the future,
Like we do for meditative states,
Or hypnosis, or the placebo effect,
As we unveil the many mysteries in neuroscience.

So I will maintain that we are the product
Of our past experiences
Run over our inherited machinery,
And each and every moment
Of memory formation and recall,
Introspective reflection, and other processes,
Are intricately and inextricably linked
And dependent upon one another.
It follows that the introspective experience
Cannot be unbiased or disentangled.

At what point can we know
That sensory perception
And information is shut off
If we are not subjectively managing this process,
As these processes are largely carried out
Below the level of explicit subjective awareness
In subcortical regions?

It may seem shut off to us,
But it is not shut off.
We may find ourselves in a state
In which we believe our awareness
Is devoid of content,
But this belief is not backed up
By how I understand the brain to work.
There is a vast amount of evidence
That our experiences are influenced
By factors we are completely unaware
(Everything from past experience,

Reward pathways, priming, among other things),
Which suggests that there may be
Influences present during such a state.
To get to the bottom of this,
We would also need to
Define information and content.

Of course, experience is subjective,
And the argument is
That common experience suggests objectivity.

But we are going to get caught up
In brain science here as explanation—
Common experience will be explained
Via commonality across brain structure
And the fact that we all
Process information the same way,
Via synaptic connections or calcium ion channels
In the case of glial cells (astrocytes).

Achieving a state that we find rewarding
Generally entails the mesolimbic dopamine system
In the basal ganglia.

This system operates under the level
Of subjective self awareness.
In this scenario, those who achieve such a state
Are then more likely to again achieve
The state because it becomes reward encoded.

If we are aware but have no content,
Information, or perception,
I'd say we are not aware.
The feeling of being connected
To something entails feeling,
Which is associated with various things,
Be it God or the universe, or something else.

Those associations are most likely processed
In part in subcortical regions,
Like the amygdala and basal ganglia.

Mystical Experience

"Mystical experience is correlated with brain states,
But it is not an awareness of a delusion,
Because delusions are incorrect
Sensory representations, and introvertive mysticism
Is specifically contentless."

I want to make sure I understand this claim—
It seems to be saying that a religious experience
In this context is not something
Which arises from a sensory stimulation
At the time of the mystical sensation?

If so then the introvertive mysticism
Is still not contentless.
This would assume that while sleeping
Or not engaging in activity means
That the brain is not processing information
(The sum total of all
Our experiences across networks
And structures which hold
Our mental models of the world,
And our biases, of course).

There is no such thing as a resting brain.
Much activity in the implicit systems
Do not make their way up
To explicit subjective awareness,
But nonetheless processing is still ongoing
And can cause sensations
In the absence of direct stimuli.

Mystical experiences are brain states.
These states can be induced,
Or reproduced by stimulation
Much in the same way that they can occur
With minimal sensory perception.

But one point we must not ignore—
We cannot turn off our sensory neurons
(In the absence of

Certain pharma products or cell death),
So even while resting or tuning out
We are still receiving signals,
Even if they are less abundant
Or robust than when we proactively perceive.

More importantly, the brain has accumulated
An immense amount of information,
Some of which causes perception
Without the stimulus—
For example, the sense of smell
Is the most powerful sense
With respect to emotional association.

However when one thinks about the smell
(Even though the smell is absent)
One can perceive the emotional association.
I would characterize this as introvertive.

It occurs frequently and there is no reason
To assume certain of these perceptions
Have any more 'standing' than any others.

AETHER?

About electromagnetic waves—
Let's think of space anywhere:

A changing electric field produces
A magnetic field and then that produces
A further electric field and so on,
This being a self-renewing disturbance,
That is, what we know as visible light, x-rays,
Infrared and so much more that are
Of this electromagnetic spectrum.

Thus, no aether is required as a medium
For electromagnetic waves to propagate
From here to there.

GENERATIONS OF UNIVERSES

From Bob Zannelli

The idea that Universes beget Universes
Has many attractive features.

One scenario the Caroll Chen model proposes
Is that a future De Sitter Space
Is the birthing stage
For the creations of new Universes.

However, as pointed out,
Several papers by Trodden
Have called this possibility into question.

Nevertheless,
The compelling logic of this scenario,
Which avoids all the difficult problems
Of eternal inflation,
Is an incentive for attempting
To find workable models
That predict the basic paradigm,
That Universe beget Universes.

One very promising scenario
Is the pre Big Bang proposal
By Venzianio and Gasperini.
In this model base
On string theory formalism,
What General Relativity predicts
As the creation of a singularity,
String theory predicts
As the creation of Universe pairs,
A Universe and an anti Universe,
The BIVERSE.

One serious criticisms of this model
Has been raised by Thibault Damour
And Marc Henneaux who point out
That space time should behave chaotically
On the approach to the predicted singularity

Contradicting the observed regularity
Of the early Universe.

One possible solution to this problem
Is to include in the calculations
The high energy string modes
On the verge of becoming black holes
Which smooth out the behavior of space time.

This solution has been proposed
By Gasperini and independently
By Thomas Banks and Will Fichler.

All models of Black Holes contain
White Hole solutions.
But we never see White Holes in Nature.

Perhaps the reason is
That these White Hole solutions
Are in fact Big Bangs,
The creation of new Universes.

THE MIRROR MYSTERY

A mirror show us *ourselves*
Right-left reversed
Because we have to
Turn around to face it.

Normally,
We turn around vertically,
Keeping our feet on the ground.

But we can face the mirror
Standing on our head—
Then, we are upside down
In the mirror
And not left-right reversed,
So, again, the reversal
Is in the rotation
Of the object—oneself.

TIMELESS REALITY

(From Bob Zannelli)

The first piece of Mathematics we need
Is what is called an Eigenfunction.
This is simply an equation that takes the form

$$OP*S=e*S$$

Where OP is an operator
That acts on S the system state,
Usually a vector, also called the Eigenstate,
Which results in the product of an Eigenvalue e,
A scalar, and the Eigenstate.

The exact form this can take
We need not worry about here.

Another function I need to mention
Is the Hamiltonian which is
The energy state of a system.

Which is;

$$H=T+V$$

Where T is kinetic energy
And V is potential energy.
You can think of T as the energy of motion
And V the energy of position.

Now we can start.
The fundamental equation
Of Quantum Mechanics
Is the famous Schrodinger equation.
This is an Eigenfunction equation.

It takes the form:

{(-hbar^2/2*m)* del^2 + V} *[Y>= E*[Y>

Where

del is the three space derivative operator,
V is potential energy,
Y is the wave function of the system, the Eigenstate,
hbar is the reduced Planck constant
And E is energy, the Eigenvalue.

Therefore we have

H= {(-hbar^2/2*m)* del^2 + V}
The Hamiltonian operator

[Y> the Eigenstate
E the Eigen value so

H*[H>=E*[Y>

This is the time independent Schrodinger equation.

Now to bring time into this
We need to have a time dependent
Schrodinger equation.
For this we have another Eigenfunction:

{ i*hbar*pd/pdt} [Y>= E*[Y>

Where i*hbar*pd/pdt
Is the time translation operator.
Here pd means partial derivative.

This gives us

H*[Y>=E*[Y> = { i*hbar*pd/pdt} [Y>

H*[Y>={ i*hbar*pd/pdt} [Y>

This is the time dependent Schrodinger equation.

Now in Quantum Cosmology
We need to write a Schrodinger equation
for the whole Universe.

To do this we need to replace
The Standard Hamiltonian
With the Cosmological Hamiltonian and Eigenstate.

When we do this we get
The famous Wheeler De Witt equation
Which replaces an operator
Acting on a three dimension configuration space
With an operator acting over all field configurations
Of space times, called Super Mini space.
This is a difficult to write in full detail
And is very difficult to solve.

But we can write simple versions
Which give us approximate solutions.
It's thought that a Quantum Theory of Gravity
Will give an exact form and an exact solution.

So we write

H*[Y> = E*{Y>

However when we look at the whole Universe

E=0

So we have

H*[Y>= 0

We have lost time from our equation.
This is one way to express
The time problem in Quantum Gravity.

Now here I offer a rather simple solution
Which I think has merit.
Remember the time translation eigenfunction?

{ i*hbar*pd/pdt} [Y>= E*[Y>

Well there is no reason
We can't write this equation as

{ i*hbar*pd/pdt} [Y>= - E*[Y>
This is because there is no reason
Why we must assume an arrow of time
In these equations and the second equation
Is the time reversed version of the first
Which perhaps is made more obvious
By writing these equations as;

+ { i*hbar*pd/pdt} [Y>= E*[Y>

- { i*hbar*pd/pdt} [Y>= E*[Y>

Now think is terms of
The Wheeler De Witt equation.
Might we not consider this equation
As a superposition of both the time forward
And the time reversed time dependent equations?

So we have.

H(+) *[Y> + H(-)*[Y> = H*[Y> = { i*hbar*pd/pdt} [Y> -
{ i*hbar*pd/pdt} [Y>=0

H[Y>=0

An equation without time.
So based on this idea we have lost time
In the Quantum equation for the Universe
Because this wave function is a superposition
Of two Universes evolving in opposite time directions.
Or equivalently, our Universe is the result
Of a previous Universe that contracted
And tunneled into a new Universe which is ours,
Or again equivalently, our Universe tunneled
Out of pure Nothingness into the Universe
We have today and a negative energy
Twin Universe evolving backward in time.

If this is true we should see this reflected
In the Friedman equations for Inflation,
Which must be able to predict inflation
In both time directions.

And in fact this is the case.

We have

(1/a)*(d^2a/dt^2) =(8*pi*G/3c^4)*rho= Lambda/3

Which solves as

a(t)=A*exp[H*t] + B*exp[[-H*t]

Where

H= sqrt[lambda/3]

So you have four solutions.

1) +H & +t Positive Energy Universe
Inflating forward in time.

2) -H -t Negative energy Universe
Inflating backward in time.

3) + H -t Positive energy Universe
Contracting backward in time.

4) -H +t Negative Energy Universe
Contracting forward in time.

Solutions 1 and 2 are obvious physical solutions.
But what are solutions 3 and 4 all about?

Well, it turns out that String Theory
May provide an answer.
These are just what is called the T duality solutions
(Which I won't explain here)
Of the physical solution
And in terms of particle physics
Represents the supplemental particle states
Proposed by Bob Klauber and myself
As the solution to vacuum energy problem.

MATH FIRST?

Physics matches the laws of equations.
Sometimes the equations are even found
Before the reality of the law

The Yang-Mills equations that govern gluons
Were discovered before the gluons themselves.
The same with the electromagnetic waves
That were predicted by Maxwell.

The Yang-Mills equations are like
Maxwell's equations on steroids.

A specific version applies
To the real-world gluons
Of the strong 'force' interaction
Using color-charges.

You can describe the properties
Of quark and gluon fields completely,
Using the concepts alone,
Not needing samples or measurements.

Since the objects obey their equations,
The 'its' are the bits.

There are regularities in nature
That lend themselves
To math equations and geometry.

Newton and others formed mathematical laws
Of motion and gravity.

MATH BEFORE EXISTENCE?

Matter is merely mathematical:
Wave-function possibilities fall
Out of eternity's equations...
Mind will grow to encompass all.

THE VAULT OF EVERYTHING

A spirit led us onward,
Who knows how,
Toward the Library of Babel,
Which contains all the possible books
That could ever be written,
Including, for example,
Better and worse Shakespeare plays,
Brand new plays,
Books with only one word
Of difference among them,
Everyone's life story
(Even the parts not lived yet),
The Secrets of the Universe,
The true Theory of Everything,
A lot of gibberish, and so on,
As we can't imagine.

[In fact, I found this story in there,
In a short story book of mine-to-be,
So I just copied it to here.
(yes, it said that too.)]

A clear night sky of infinite possibility
Showered us with photons,
Lighting our way
To the fountain of all knowledge.

"True enlightenment awaits me there,"
I offered to the guiding spirit.

"Don't be so sure,
Although you might chance upon it,
For the deep truths of enlightenment
Are as needles surrounded and consumed
By the near infinities of the stacks
Of deception and confusion,
For, remember,
EVERYTHING exists in this library."

"It must be a massive building," I remarked.

"Well, yes, but it's bigger on the inside
Than on the outside;
Otherwise, it would have been
Larger than the universe."

"Bigger on the inside? How?"

"Well, you'll see, but I'm not sure how—
Maybe through some dimensional extensions—
Or perhaps it's constructed digitally
And expands as you move about, somehow,
To conserve space; but, even with compression,
It's still hundreds of miles wide
In every direction—on the inside."

"What is Everything, in principle?"

"Every arrangement possible,
Given whatever constraints there are, if any.
Of course, not all paths may be stable,
Sensible, or last very long."

"That's a lot—
Why do we live on this particular path
That our Universe has taken?"

"Who the heck knows!"

"What about making the forms of
Substance(s) of a Universe?"

"Well, in the case of the emission
Of the secondary substance(s), let's say,
It's every one of the 'alphabets'
That can be conceived by
The Timeless-Formless-Motionless,
Plus, all of its resultant workable combinations
And interactions of substance.
For this Babel library,
It is every possible arrangement
Of words in every language,
With punctuation, too, naturally."

"Hey, here it is. I can't wait!"

Upon entering, they saw stacks of books
In every direction, even up and down,
Stretching toward infinity.

"Where's the card catalog?"

"There can't be any,
For many titles and descriptions
Of similar books are too long to differentiate.
Think of the books themselves
As the card catalog."

"How's the library organized?"

"It can't be. It would take forever."

"Who runs it?"

"Borges is the lone librarian,
But he's somewhere in the back
And hasn't been seen for decades."

"OK, I'll pick some at random."
(Hours pass)

"Anything?"

"No, mostly mumbo-jumbo,
But I found one on a table
That someone must have treasured."

"Oh, yes, he spent his entire lifetime here.
It's Plato's 'Beyond Metaphysics'."

"Wow! That's been lost for thousands of years.
But is it the true version?"

"Who knows."

"This library contains
No information whatsoever!"

*"True, but there's another library next door
That also claims to have Everything."*

"You mean that little 'hut'
No, wait—I get it—
The library next door is empty."

"Yes, for the All sums to the None."

"Wait, I found two more good ones
In the stack right near the entrance..."

*"One is by you and one is by your friend, Rascal.
You put those there in the first stack
So someone would find them easily
And read them, even though they exist again
Somewhere else in the library."*

"Yes, and I'm even going to let them
Stick out a little on the shelf."

...

In another chilling Borges's story,
I read the actual book that he refers to,
The one whose infinite pages
Are ever-changing,
For that's how books appear to me
In my night dreams.

Sometimes there are even digits occurring
In the middle of words,
Plus, if I look away and then back,
Then the contents of the page have changed.

One time, when the page stabilized
To quite understandable words,
I realized I was reading
Something very profound.

In fact, it was the Ultimate Answer.

I dared not look away
Nor try to copy it with a dream pencil,
But, instead, tore out the page
And crumbled it into my hand,
Then forced myself awake
(it was a lucid dream).

When I awoke,
I had the page in my hand,
And it said:

This page intentionally left blank,
Except for the above,
And the above, etc.

LIFE FROM INTELLIGENCE THEORY

[theory]
...endowed with intelligence
That had to come from somewhere.

The Proposition:
Life requires intelligence behind it.

What a fine mess your theory
Has gotten us into:

It has that no further INTELLIGENCE
Was required for a really Big Life to be (God's),
But that the way lesser case
Of life on Earth fully requires it;

Thus, the theory fails right off the bat.
Flunk. Back to the drawing board.
Not QED.

NO CREATION

It was impossible for there
Not to be “something”,
Or the capability thereof
(Quantum tunneling),
Which is still “something”.

So, then there would be no Creator,
Nor none even necessary.

This is because Nothing
Could not produce “something”,
Nor anything,
As Nothing has no properties,
Capabilities, etc.;
So, “something” had to be, plus it is,
Plus, a lack of anything (Nothing) is not.

No Divine plan has been seen or is required.

Consciousness is obviously tied to the brain.
Anything talking to the brain
Would have to speak the same language;
Thus, no immaterial stuff of a “soul”.

OBEs and NDEs are totally
Explainable by science.

There is no proof of past lives whatsoever.

Morality was/is provided
By humanity / natural selection.
The afterlife is but a wish.

Being “special” is not the ultimate freedom;
No strings attached is the ultimate freedom to be.
The wish is not even ever so humble,
But ever hopeful and even greedy for reward.

THE PRIME DIRECTION

...Order to bring harmony
Within the human-world-system.

The music closest to the TOE
Is the harmony of the TOE itself,
A symphonic orchestra consisting
Of the universal, galactic,
And solar sections
All playing in concert;

Yet, all their instruments are still separate,
No one sound rising above the others,
Such as the pattern between the patterns
Arising to make the prime number keys,
Much like a string plucks its harmonics
Of 1/2, 1/3, 1/4, etc.

It's not quite the music of the spheres,
Yet, they, too, resonate to it,
Flinging it down from
The Father sky
To our Mother Earth.

So, the songs of life, too, are sung to it,
Yet they are vibrations of it,
And all music repeats it.

It is the pattern outside of the patterns,
The primes conducting all the rest
Of the musical numbers.

(The proof may remain incomplete
Even into the Year 1,000,000,
Yet it remains "conditionally" true.)

Head, Math, and Beyond

Of the Seven Wonders
Of the Old World,
Only the Giza pyramid remains,
The rest having succumbed
To fire and earthquakes.

The life of some things approaches forever.

The prime numbers march on,
Never ending, as we will see,
Although the non-primes
Ever chip away at the prime real estate.

Alien beings from Vega
Beam a series of prime numbers to Earth...

Primes are the atomic elements of arithmetic,
From which all the non-primes can be formed.

What is the secret of their pattern?

It would seem that only the non-primes
Have a pattern, those being their nth instances.

How then can can there be an unobviously
Unpatterned pattern between the patterns
In these leftover spaces in between?

The even numbers already cut
The prime potential in half,
But for '2',
Which is the only even prime.

The multiples of '3' then
Remove another swath,
Although less than a third,
'4' doing nothing,
And '5', some more,
And so on,

These few numbers alone
Already consuming
Perhaps over 70%
Of the numerical realm.

Yet, there will ever be more primes,
Some even so-called "twins",
Like 1,000,000,009,649
And 1,000,000,009,651,
For, the harmonics may only
Approach 100% but never get to it.

Bertrand Russell once wrote and thought that
"Mathematics possesses not only truth,
But supreme beauty—
A beauty cold and austere,
Like that of sculpture,
Without appeal to any part
Of our weaker nature,
Without the gorgeous trappings
Of painting or music,
Yet sublimely pure,
And capable of a stern perfection
Such as only the greatest art can show."

However, in later life,
Bertrand dismissed his youthful rhapsodizing
As "largely nonsense", writing
"Mathematics has ceased to seem to me
Nonhuman in its subject matter.
I have come to believe,
Though very reluctantly,
That it consists of tautologies."

Will math become even
More trivial than six of the Seven Wonders?

Primes seem to be special since
They cannot be split into smaller factors.

All the unspecial non-primes can be obtained
By multiplying primes together,

For example,
'666' is 2 x 3 x 3 x 37,
This being called
The "fundamental theorem of arithmetic".

How many primes are there?

There are an "infinite" number, as Euclid shows,
For, if they were a finite number,
Then one could always obtain another
By multiplying the primes together and adding '1'.

What then the pattern of the "primal" scattering,
Seeming more as random weeds sprouting?

Are we in the presence of one of
The inexplicable secrets of creation?

Are they of a complex and timeless reality
That is independent of our minds?

Are they transcendently mysterious?

No, for they obey a law.

The Law of Primes

Here we must climb the edifice
Of mathematics that rises above
The humble counting numbers,
The fractions, and the real numbers—
All the way up to the
Complex numbers with "imaginary" parts.

Here we find the Riemann zeta hypothesis
That holds the secrets of the primes,
After an ascent taking over two millennia.

If true, then there is
A hidden harmony to the primes,
One that is rather beautiful.

In 1900, David Hilbert included it
On his list of the 23 most important
Problems in mathematics.

It is the only one that remains.

Computers have thrown
Zillions of numbers at it
And yet it holds, never failing;
But, what would it do further on
In its path toward "infinity"?

If it ever fails, even once,
Then all the thousands
Of theorems of higher mathematics
"Conditioned" on its hypothesis
Will collapse into a heap of meaninglessness.

What, then, and where,
Are the ever shrinking recesses
In which primes can grow forever?

The zeta functions has its origins in music,
The vibrating violin string
Creating all the overtones of the note,
Know as the "harmonic series"
Of 1/2 + 1/3 + 1/4 +...

If we take every term in this series
And raise it to the variable power 's',
We get the zeta function,
Introduced by Euler, around 1740:

zeta(s) = (1/2)**s + (1/3)**s +, etc.

He then noted that this infinite sum
Could be rewritten as an infinite product:

zeta(s) = 1/(1-1/2**s) x 1/(1-1/3**s) x ...

However, he didn't fully grasp
The potential of his

Infinite product formula,
Writing that the human mind
Would never penetrate
The mystery of the sequence
Of the prime numbers.

In 1792, Gauss noted that one could estimate
How many primes there were up to a given number
By dividing that number by its natural logarithms,
The percentage of error heading toward zero
As the number got larger.

This, then, was
"The coin that nature tossed
To choose the primes."

Again, it might not hold unto infinity,
Nor does it predict any primes.

It was Reimann who would dispel
Any lingering illusion of randomness/mystery.

The Prime Hypothesis

In 1859, Reimann cracked the mystery.

He began with the zeta function,
Enlarging it to take in
The complex numbers,
Those having both
A real and an "imaginary" part
Involving 'i', the square root of '-1'.

The complex numbers
Are two-dimensional,
So they can be graphed on a plane.

In effect, Reimann created
A vast imaginary zeta landscape,
Consisting of mountains, hills, and valleys
That stretched forever in every direction.

The sea-level points,
Those with zero altitude
That yielded the zeta output zero,
Were the most interesting,
For they showed exactly how
The infinity of primes
Arranged themselves
In the number sequence.

There was no longer
Random noise in the primes,
For now there was
A way to hear their music.

Each zero of the zeta function,
When plugged into
Reimann's prime formula,
Produced a wave resembling
A pure musical tone.

When these pure tones are all combined,
They reproduce the structure
Of the prime numbers,
The particular location of any given zero
In the zeta landscape determining
The pitch and volume of
Its corresponding musical note,
And, very importantly,
The farther east it was,
The greater the loudness.

With all the zeroes lying in
A fairly narrow longitudinal strip
Of the zeta landscape, and only if,
Can the orchestra of
The primes play in balance,
With no instrument
Drowning out the others.

But Reimann went further.

After navigating just a tiny portion
Of the infinite zeta landscape
He asserted that all of its zeroes
Were precisely arrayed along
A "critical line" running
From north to south—
And it is this claim that became known
As the Reimann zeta hypothesis.

Epilog

The ebb and flow of the primes, then,
Is the pattern of each instrument playing,
But combining together
With the others so perfectly
That the patterns cancel themselves out.

We may predict that long before
The year 1,000,000 A.D.,
Mathematicians will awaken from
Their collective Platonist dream,
Noting, like Bertrand,
"You are only Symbolic Convenience."

While even the Giza pyramid may crumble,
Along with the magic of numbers,
We will still have laughter—
The so-called hardest problem
Of the primes then becoming
A somewhat broad joke
Of a trivial tautology
To the schoolchildren
Of that distant time.

WHAT IS LAUGHTER?

Hardly anything is deemed
To be more parochial
And ephemeral than laughter.

Or more lowly,
For, during much of human history,
The comical has been a mix
Of lewdness, aggression, and mockery.

It begets a peculiar panting
And chest-heaving behavior,
Traditionally viewed as a
"Luxury reflex" serving no purpose.

However, could it be
That it derived from the "false alarm"?

A seemingly threatening situation presents itself;
You go into the "flight or fight" response;
However, the threat proves spurious,
So you alert your social group
To the absence of real danger
By emitting the vocalization of laughter,
One that, as it passes,
Is amplified, contagiously,
From member to member.

The mechanism was then hijacked
For other purposes such as
Hostility or showing superiority.

Still, at the heart of laughter
Lies incongruity,
Such as a grave threat
Revealing itself to be trivial.

It has become a kind of intellectual emotion,
Every joke being an interrupted syllogism,
For example,

"The important things is sincerity.
If you can fake it, you've got it made".

The odd and incomprehensible
Suddenly turns into nothing.

TAO?

To me the Tao,
And I'll just speak of the Tao now (not God),
Is eternal.

The yin/ yang flow is eternal,
The self made boundaries is eternal,
Its dynamic symmetry is eternal,
And it's all part of you and I and the universe.

This explains our physical dynamics
But it's not very good at explaining
Our internal dynamics of emotion,
Imagination, intelligence, etc.

Those type of changes are more applicable
To the internal dynamics of the the I Ching Trigrams.

Good stuff, that old TAO,
Although there is still some embroidery
That some will always stitch upon it.

Basically, the ground-state must be eternal,
And therefore causeless of any specific intent,
Its emissions having symmetry
So as to have energy be conserved
By being neither created nor destroyed.
And, true, we are all of it.

Our internal dynamics come later on,
Emerging from the complexity that in turn
Was made of the elemental units.

TO THE DEEP

To learn the Secrets—what IS and ev'r WAS,
We must brave the crypt and ghost of cause...

So, into the deep, we go, without pause,
To look down, ever down, no self to keep—
Through birth, death, and the shade of sleep,
Through paths unkempt, underswept—to the deep,

Through the cloudy strife
Of this hazy life,

Through the equations of eternity,
Their non-paternity nor maternity,

Past the realm of the things which seem or are,
Even o'er the steps of the remotest bar.

Down, down!
Where the mind whirls round and round,
As the ear draws forth the sound,

As the eye sees the light,
And of the dark the fright.

Down, down,
Beyond all death, despair, love, and sorrow,
Past yesterday, today, and tomorrow,
The body's guide is but the logic of the mind.

Down through the fog, the not, and the void,
Where "God" and Nothing fail; Oh, zoids!

Down! Where reigns the night, and the air is thin,
To where the sky and stars are not, but within,

Where the radiant have not their throne,
Where there are one or some pervading, all alone.

Down, down! To the fathoms of the cryptic;
Down, down! Where substance slept with arithmetic,

Toward the spark yet nursed by embers,
To the first and last that the universe remembers,

To seek the gem that shines—the wealth of mines,
The jewels so treasured by thee and thine.

What accelerated life's momentous gem,
Letting your motto be "Carpe diem"?

What seized the moment or lost its momentum?
Wearing not the time as its royal diadem?

World does not pass by—we pass through it;
Clear your being so the treasure may arrive;

This spirit sparkles of a different light—
The gemstones are of a different mine.

Down, down! We guide thee, we must carry thee;
Down, down! We're illumination beside thee...

Fear not the proof—it's the beauty of truth:

What entropic seas never-endingly
Cast Deathly Time aside,
Ceaselessly thriving on...

Of that which was the imperishable—
The flame of beauty inextinguishable,
Forever celebrated as immutable...

That gained its perpetual permanence
From the undying love of the glorious truth?

As, once, above the ground, you're born again
When the roseate hearts were cleansed by dew;

And lucky were you if spring found you new,
As every blossom on the bush blew full;

When these wonders the new morning bestrew;
The beauty of truth was all that you "knew"...

For, life's hardships were softened by beauty,
All its weaknesses strengthened by truth.

As when roses blossomed, like realizations,
Beauty itself bloomed from the well of truth.

Down, down!
For now, rarely enough, existence is left aside,
And, yet, the essence has its other side—

Life, although anguishing, must be lived fully,
Since, if we're alive enough to feel its beauty...

Then we're exposed to the opposite twin—
Yes, Beauty's other side is Melancholy.

Down, down, the essence beckons us home,
As like the contained-container is the poem.

When a deep truth is known so intensely
That all of its clothing falls away...

Then we have learned the beauty of truth,
For the reality of meaning is beauty.

Opposite twins rule the causing call,
The positive and negatives being all.

When sadness brooded over the morrow,
One visited the deep well of sorrow.

There enshrined, inseparate, Beauty said,
"'Twas from me that sadness you borrowed."

Do we live the life of art,
Each playing our part?

Nay, that is not life, nor a part, bit,
For there's another dimension to it.

Art and poetry enrich human experience,
But they're no substitutes for the living of it.

Like Keat's figures on the urn, should we live life less?
NO!—because what is deathless is also lifeless!

Down, down!
Truth and beauty must be inseparable,
Although seemingly imponderable.

On that sphere above,
Soft breezes blew, caressing me and you,
As we kissed the roses and drank their dew.

Reason and passion merged into one,
As truth and beauty made their rendezvous.

Down, down,
Through all antiquity, past all of the known
Arriving at the lowest, remotest throne...

But not one of the highest perfection,
For that will become of the opposite direction.

Here, the enigma of the immortal
Is undone and unloosed, through its portal—

The Theory of Everything mortal—
An Idea for which we've opened the door to.

Down, down, to the end at last!

Here the timeless, lawless, and formless
Of the unordered, uncreated scene;

Here the causeless reigns "supreme".

"IT" FROM BIT

Information IS Reality?

A quantum entity can remain
In an indefinite superposition
Of an "everywhere and nowhere" indefinitely,
Since, perhaps, the expenditure
Of gravitational energy is so negligible
That it doesn't "matter".

At macroscopic levels, such as with us,
Collapse to a definite place
Occurs in a zillionth of a second.

Somewhere in between the tiny and the large,
Say, something like a spec of dust,
Collapse may even take a second or so
For gravity to collapse it to one place.

That was introductory,
So, now, what is "it" from bit
Exactly, for a quantum entity?

A "bit" is what we have
When we gain information
About a quantum entity,
Such as its location or momentum
From an observation, a mark,
Or some recording.

This is called "registration",
Whether done by a person,
A device, or a piece of mica;
In other words.
Anything that can preserve a record.

Only then does the quantum entity
Become an actual "it";
So, "it" comes from "bit",
Which is information.

Until then, the "quantum entity"
Was not yet an "it",
As there was no objective reality gained.

The laws of quantum physics, then,
Only tell us what may happen;
While a measurement tells us
What is happening (or what did happen).

Perhaps, then, it is that
Information sits at the core of physics.

The total universe would be the big "It"
That arises from the myriad yes-no choices
Of measurement (the "bits").

So, it would be that information underlies reality;
However, this information is not just
What we learn about the world;
It is what makes the world!

An "it" from bit reality example:
When a photon is absorbed,
And thereby "measured",
An unsplittable bit of information
Is added to what we know about the world,
While, at the same time,
It creates the reality
Of the place and time
Of that photon's interaction.

Before its absorption,
That photon had no true reality.

The universe would seem
To be made out of discrete quanta.

Another example:
Concentrating on their spins,
A two-electron system contains two bits.

For example, they might be
“The spins in the z direction are parallel,”
And “The spins in the x direction are antiparallel”.

The two bits are thereby used up,
And the state is completely described;
Yet, no statement is made
About the direction of spin
Of one electron or the other.

The entire description consists
Of relative statements, or correlations.

This means that as soon as one spin
Is measured along a certain direction,
The other one is fixed,
Even if it happens to be far away.

Thus, quantum entangled particles do not have
Pre-existing properties, such as polarization,
That are independent of any observation.

This is the fall of naive realism
[At that level].

The result is so random
That not even God could know the answer.

Thus, randomness is ultimately
A consequence of the finiteness
Of the information.

A quantum system can carry only
A limited amount of information,
Which is sufficient only
For a single measurement.

Two particles collide,
And in so doing
Enter a state of limitation.

In terms of information theory that means
That after the collision the entire information
Is smeared over both particles,
Rather than the individual particles
Carrying the information.

And that means the entire information we have
Pertains to the relationship
Between both particles.

For that reason, by measuring the first particle
We can anticipate the speed of the second.
But the speed of the first particle
Is entirely random.

Quantum information is
Such that a bit can be 0 or 1;
A measured particle ends up
Either here or there.

But if a particle carries only
That one bit of information,
It will have none left over
To specify its location
Before the measurement,
Because the information was not sufficient.

Randomness is reality's bedrock.

What about further on up.

It would seem, then, that,
Somehow, "less is more",
But also that "more is different".

When elementary units
Are put together,
It is that we get something
That is more than
The sum of the units.

For instance,
A substance consists
Of many molecules,
Gaining properties of
Temperature and pressure
That no one molecule has.

It, too, may be a
Solid or a liquid or a gas,
Yet, no one molecule
Is solid or liquid or gas.

When enough simple elements
Are stirred together,
There is hardly any limit
To what can result;

Thus, the complexity of the universe
As a whole does not preclude
An extremely simple element
Such as a bit of information
From being what
The universe is made of.

Does "more is different", then,
Have something to do
With "it from bit"?

(Gleaned from various readings
Of Penrose, Anton Zeillinger,
And John Wheeler)

INFINITE REGRESSES, THE KISS OF DOOM UPON A THEORY

Who made the Laws and endowed them
With what must be intelligence
To follow the Laws?

When positing intelligence
Such as God's Intelligence
As a must for the universe,
The intelligent plants,
And the intelligent human life in it,
We must then carry on this thought
For it to be correct,
Not stopping and suddenly changing it
By quickly obviating it,
But even ever moreso then requiring
A further INTELLIGENCE
To explain God's Intelligence,
And so on,
This required infinite regress
Thus never ending.

That's why it's not an answer
But only a larger question
Rendering the whole premise
And its "answer" beyond repair.

When one makes a general requirement
For a cause to be Intelligence
To explain intelligence
One must totally stick with it
If one wants to be consistent,
Lest one has only scratched the surface
And gone away, stopping prematurely.

Also, look up how and why
The sunflower's sunny face of "seeds"
Are arranged in the Fibonacci sequence.

Science ever uncovers
What was formerly thought to be "magic".

ENIGMATIC MATHEMATICS

My general impression is that
The enigmatics of mathematics
Mainly points at the enormous
Richness of mathematics.

Could be just that, which is indeed a lot,
Or it could be that something
Even larger—information
Is the basis for the "its"
Being the same as the "bits",
Such as in the Yang-Mills equations
That so much describe the gluon behavior
That we don't even have to try
To experimentally look at them
To gain the same empirical information.

Amazingly, the equations even came first,
Before the gluons' discovery!

I gave a link in which Michio Kaku
Explains Einstein's method of thinking:
How did God invent this universe?

And, well, if you follow that, one would say:
Let's first start with inventing
The thinking tools that are rich enough
For the kind of universe I have in mind.

Hence, mathematics came to be.
Probably the kind of mathematics
That is even richer than we already know.

That's sounds good,
As there is much more math to discover
(We can't even solve the 3-body problem),
But then there is the same
And further problem for God's Life
Being similarly prepared for by (?)
If we carry this theory through.

We may find richer math in the future,
Pushing Godel's boundary further out.
One can win one million dollars
By proving the Reimann zeta hypothesis.

In his eternal question Einstein went further:
Did God have a choice?

In other words:
What was there first?
What was there being stubbornly eternal?
What was/is there, I think, being TOE?

All we may ever find as a description
Of the [Super]TOE,
Although still a complete description,
Is that the ground-state
Had to be without cause,
And thus of no real order
Not imparted to it, since nothing prior.
I think we have to understand
That TOE is really very fundamental.

Yes, and probably quite uninteresting
At its source since it must be a simplistic state.

As those states are always "unstable",
As simple things are,
They can go through phase changes
Into the more complex states, on and on.

And in a very general way
It concerns quarks and the Milky Way
And the Elephant and everything
In between and around it,
Being universe, being, whatever, multiverse.

While TOE also is within our though
And world view and culture
Because we are part of this TOE-system.

Therefore, thinking TOE is a kind of trying
To align ourselves with this TOE-system-universe.

True, TOE must also include
The explanation of all
The emerging complexities that followed,
Right on up to how we ourselves operate.

We will find much that is new,
Yet forever mixed with what is old,
Such as appearances in motion
From Then to When,
Granting the past-present-future,
And a Where and a What
Of events moving on and changing
In the place of a space.

The Why and the How of the initial TOE,
Although kind of resolved
As Nothing not being able to be,
And the positive/negative
Particle pair emissions,
Is an essence that,
While great for curiosity
And some deep meanings
Of what we really are,
Is still dwarfed and trumped
By the glory and state
Of our everyday existence that is life,
Totally overwhelming it.

And all this is based on very old mythology.
It's a promise, a hope, of coming home.
Though we won't know it for sure
When something TOE pops into public
Because we don't have
An absolute view on reality,
But it would/will help.

In getting to grips with these
Other problems,

Of surviving ourselves on this planet,
And being universe with this universe.

Right, it's the present and the future that counts,
Our minds thence progressing higher.

We've been looking for "higher"
In the complete wrong direction,
For "higher" was not in
The small and simple past,
But is yet to be in
The further emerging complexity
Of the future.

Onward!

As it seems that all that is possible
Must eventually happen somewhere,
Our universe must have bubbled up
In such as way of harmony that we,
Of course, must only find ourselves there,
And a lucky thing it is, too, that we are,
Given such events as asteroids
Wiping out most life gave us a clearing,
As well as two monkey chromosomes
Fusing gave divergence
To our homo sapiens life.

If these didn't happen,
Something similar may have
In some other bubble.

It's still fun to explore it,
Even knowing that it
Must always mesh.

THE THREAT

The Taliban's top leader is in custody.
Mullah Abdul Ghani Baradar
Was nabbed last week in Karachi
By American and Pakistani intelligence teams,
Unnamed U.S. officials told several news outlets.
Two Taliban figures also confirmed the arrest.

Since the US has signed the Geneva Convention,
Which prevents the torture of prisoners
To gain valuable information
Which would save many lives,
(Although the Taliban haven't signed it)
The NIA was called in
To question the prisoners.

They were able to find out
About some nuclear material
Being brought in to the US;
However, the cell involved is self-contained
And so there'd be no way to contact them.

Meanwhile, Analog,
The southern field commander
Of the Ninja Empire,
Had been alerted by TimeParticle
Of the "unusual event" of the sinking
Of a Coast Guard drug-interdiction vessel
Off the coast of Texas.

Timeparticle:
"It's not often that these vessels go down;
Either it was sunk because
Of a huge amount of drugs being involved
Or because something much more potent
Is being brought into the U.S."

Analog:
"Let's try our new satellite system."

Timeparticle:
"It has photos better than Google Earth?"

Analog:
"It has super high definition video
That gets saved for but a few days
On holographic discs."

TimeParticle:
"Of the whole gulf and the southern region?"

Analog:
"Yes, amazingly, it does,
But that's a lot of data and so that's why
We have to reuse the storage after a few days."

TimeParticle:
"That is amazing,
Having every square inch
Of such a massive area on HD video.
Let's take a look."

P: I'm sending a feed to the General.
Let's start it a few minutes before
The Coast Guard vessel went under.

TP: Wow! This is more than just HD video—
It's 3-D+! And there are no clouds.

P: We have two satellites, to provide 3-D;
They can also take away the clouds
And then enhance the scene
From the infrared and more,
But this is a clear day.

TP: The Coast Guard vessel
Is hailing two speedboats and they're coming to a stop.

I can't quite understand what's happening
Inside the speedboats
From this topdown view.

P: Change to more of a side view.

TP: We can do that? OK, done.

TP: Uh, oh. A handheld missile launcher
Is being aimed at the Coast Guard vessel.

TP: They sunk it,
And are now taking off.

TP: Cripes! They must be doing 100 mph.

P: Run it fast forward;
We'll track them to their landing.

TP: OK.

A (to comm assistant): Get me Oversight,
And control of the laser satellite.

Comm: OK; ...
The laser satellite is presently
Over the other side of the world.

A: Bad luck. Get me a live Stealth Bomber.

Comm: Yes, sir—one is already aloft
In the region; it's even NIA.

A: Good luck. What's up, TP?

TP: See, they're at the Texas shore now.
I've slowed the video.
They're loading something into a white van.

A: Good work.
Pinpoint that van and follow it into real time.

Comm: Niihau is on the line;
They've been following;
They are granting you
Ultimate Authority to continue.

TP: I've tracked them using high speed video.
We now have real time, via the NIA Eye satellite.
They've entered Houston proper.

A: Holy Christ!
I need a hundred cement trucks.
Alert Houston police pursuit,
But have them stay back, slowing all traffic;
The terrorists might get scared
And detonate and we're not quite ready yet.
Redirect the Stealth jet, highest priority.
Have the counter-terrorism official
At the next toll booth
Take a special reading on that van.
Get me the U.S. President.

...

Meanwhile, ironically, at Capitol Hill,
The politicians are again raising a fuss
About the severe prisoner interrogation techniques.

...

TP: The van ran the toll booth,
But we got a reading that
The nuclear device is live and armed.

P: Damn! That means they are
Approaching their target very soon.

TP: NASA?

A: It would appear to be.
Evacuate the area to the east first.
How's our jet doing?

Comm: They're in the area,
Having just pulled out of mach-4;
Just about ready;
Here's a direct connection to the pilot.

A (to Pilot and Navigator): Emergency verified.
Do not hit the van directly;

The device is armed;
Blast some bunker-buster holes
Just ahead of the van
Such that it will plunge into one of them.
Execute at will.

Comm: The cement trucks are on the way.
General Rascal likes your plan.

P: Have the police pursuit
Stop and block all lanes
To stop all the traffic behind the van.
The traffic ahead of it will be able
To just move on in time.

Comm: I have the U.S. President.

A (to President): NIA request to prepare
For DEFCON 3, as we may very soon
Have an unstoppable nuclear explosion
Within Houston city limits.

The President: Will do; have been following.

The Stealth Bomber bore down
Upon the speeding van,
Like a giant bird,
Blasting three holes in the freeway
Just in front of it,
The van plunging into the first one.

The jet sent more missiles near by,
Some of which might cause some debris
To partly fill the hole the van was in.

A: Nuclear explosion?

TP: No.

P: They were still three minutes from NASA.
Keep the area clear.
...

TP: We have an underground
Nuclear blast confirmed.
Low megatons.
The wind is still out of the west
There's not much population there.

A: Ah, good luck!

TP: Indeed.

A: This should end some of those discussions in D.C.

TP: Let's backtrack those speedboats
Whence they came.

A: Get on it.

MORE INEFFABLE EXAMPLES OF EVOLUTION IN ACTION

Trees grow constantly taller
At enormous evolutionary *cost*
In their competitive fight for sunlight.

Ichneumon wasps lay their eggs
In the paralyzed yet still living bodies
Of caterpillars so that the new larvae
Have instant access to a *sustainable* food supply.

Dawkins does not rail against
The book of Genesis so much anymore.
He doesn't need to.
He need only explain Darwin's theory
Of incremental steps trudging
Mindlessly through time.

THE LIGHTNESS OF BEING

Does anyone wonder why such smallness
Is at the bottom of all that later gets writ larger?

Is it that there is some great power
Of flexibility in the really tiny elementals
That larger fundamentals could never possess?

Now, wait, isn't size relative,
Such as many things are,
An atom possibly being an universe
Or or our universe really an atom,
As is repeatedly "discovered"
In late night dorm room discussions?

No, for the scale of size is absolute,
For the Planck size is near the limit
Of anything going inwardly small,
Of how tiny something can be;
Thus, tiny is indeed small,
Even miniscule, absolutely.

And, although, theoretically,
There is no limit to
How large a structure can be,
It may well before that collapse into itself.

Somehow,
The elementals being very tiny
Seems to be a key
To all that comes forth,
As we could see
That bigger, cloddier elemental things
Might have too much gravity, stability,
Or something else that would limit them
To do little more than accumulate [or not]
And make a black hole or whatever
Doesn't really go anywhere
Into complex composites
Of recombination.

Now, before we run away with this thought,
It should also be, perhaps,
That it, it must be, as well,
That the emitting puffs of energy are small,
Thus, the sizes of the elemental particles are
Reflected in that which may have been a limit
Of what the vacuum energy could do.

MIRRORING TO EXTREME

"Imagine these same circuits
(mirror neurons and
And frontal inhibitory structures)
Become *hyper*active
As sometimes happens when you have
Seizures originating in the temporal lobes
(TLE or *temporal lobe epilepsy*).

The result would be
An intense heightening
Of the patient's sensory appreciation
Of the world and intense empathy
For all beings to the extent
Of seeing no barriers
Between himself and the cosmos—
The basis of religious
And mystical experiences.

(You lose all selfishness
And become one with God.)

Indeed many of history's
Great religious leaders
Have had TLE.

The late Francis Crick,
Has suggested that TLE patients
As well as priests may have
Certain abnormal transmitters
In their brains that he calls "theotoxins".

THE LARYNGEAL NERVE

Let us untangle the recurrent laryngeal nerves.
Two of the many cranial nerves begin
At the cranial stem and head toward
Their final destination in the larynx.

One of them, as we would expect,
Takes the short direct route.
The other takes
A seemingly ludicrous excursion south,
Through the chest cavity,
Below the heart, makes a u-turn,
And comes right back up
To terminate at the larynx.

Dawkins remarks,
"If you think of it as the product of design,
The recurrent laryngeal nerve is a disgrace."

If however one examines this phenomenon
With *evolutionary* eyes,
We learn that our aquatic ancestor's
Embryonic brachial arches formed gills
As well as the ventral aorta.

As these features evolved,
The connections were maneuvered
To facilitate new features.

Instead of altering the laryngeal nerve path
To a more direct route
(Thereby avoiding disgrace),
It was pushed aside during the transition
And there it remained.

What we're left with today is our own
Comically circuitous laryngeal nerve route.
The laryngeal nerve is
An example of history, not design.

THE CRUCIAL LENSKI EXPERIMENTS.

Bacteriologist Richard Lenski
And his colleagues at Michigan State University
Have been conducting experiments that are
A "beautiful demonstration of evolution in action."

Beginning in 1988, the Lenski team has followed
The evolutionary lineages of 12 separate populations
Of the bacterium *Escherichia coli.*

For bacteria, generations are measured
In hours or even minutes,
Making them the perfect organism
For evolutionary studies.

These generational flasks of E coli
Also contain a carefully controlled brew of nutrients,
Primarily glucose, providing researchers the ability
To *tinker* with the population's capacity
To process its food.

Right around 33,000 generations,
One of the 12 lineages suddenly exploded
In population density by more than six fold
Over the other 11 lineages.

To use Dawkins' own vernacular,
This one lineage "suddenly went berserk."

This sudden dramatic growth was astonishing,
And the explanation for it is breathtaking.

Although the "broth" in each flask
Was primarily glucose,
It contained other nutrients as well,
One of which was citrate.

But *E. coli* cannot process citrate as food,
U*nless it mutates*, which is exactly what it did.

It changed the rules.
It developed the ability to eat citrate.
It *evolved.*

This lineage, from so simple a beginning,
Had had enough of the restricted glucose diet,
And *figured out* how to *eat* citrate!

NOTHING AT ALL

Don't you think nothing might be
An initial precondition which holds all possibilities,
All potentialities, all undifferentiated opposites
In a state outside human experience
Beyond space and time?
Couldn't it be pre-existence
Holding all possibilities for existence?

This could be close, in a way,
Although Nothing has zero properties,
Being a total lack of anything,
Even of a potential/capability
To differentiate "itself";

But, to ignore that and go forth
On this thought of sorts,
One would think that Nothing,
Being the most perfect and simplest state,
If it even tried to form or dominate,
Fluctuate, as does the quantum realm
Into a state of loose and jiggling change,
From which then all possibilities could happen,
Putting forth all those opposite particle pairs
In all kinds of various ways,
Some stillborn and some not...

Our little bubble of a universe
Obviously, would be one that functioned
To the extent that life could form.

SMART PILLS...

The DNA guy, Watson, may be getting, um, old,
For he is recommending genetic changes

While there are no approved smart pills,
There are those taken by college students
To stay up all night,
Plus there is always good nutrition.

However, consider cigarettes,
Which allow the brain to better focus.

Indeed, why pay good money
For a modern concentration camp
When the same can be had from smoking
For even more money.

Do not worry about any disadvantages
Such as death.

Those smokers dying young
Just didn't keep it up long enough.
Even George Burns could have
Kept on going smoking cigars,
But he quit just before he died.

I HAVE IT—THE TOE

The beginning was so simple,
As it must be at that point;
So, I've also detailed the first few instants,
Along with all of history,
Even taking it into the future.

I'll put it for free at the end of this book
If someone posts an image of a fine looking lady.

THE BIRTH OF MODERN COSMOLOGY

In 1877, when Mars
Approached Earth very closely.
An italian astronomer
Observed some dark markings,
Calling them "canali", in italian,
Meaning "channels".

Percival Lowell was later intrigued
By the find of these "canali",
The word perhaps taken by him
To mean canals,
For some were straight,
Implying that they had to
Be made by Martians.

It would be, of course,
That Mars is a dry planet
And so water would have
To be transported
To the more arid regions.

Strangely, all this led to one of
The strangest discoveries in science:
That most of the universe was missing.

Percival Lowell had by now developed
An obsession with life on Mars.

In 1894, he decided that
The clear Arizona skies above Flagstaff
Would be perfect for the task
Of finding this alien civilization.

Vesto Melvin Slipher, an Indiana farm boy,
Became Lowell's assistant in 1901—
Taken on, reluctantly, by Lowell,
As a favor to a friend,
For a short fixed term.

Slipher left 53 years later,
When he retired from the position
Of observatory director,
Having kick-started modern cosmology.

Spiral nebulae, called "Island Universes"
By Immanuneul Kant,
Were an enigma at the time.

Slipher had been using the Clark telescope
To measure whether the nebulae
Were moving relative to Earth,
Utilizing a spectrograph,
An instrument that splits light
Into its constituent colors,
Realizing that the colors would change
If the nebulae were moving
Towards or away from Earth,
Such as if a rainbow moves away from us
There will be a resultant red shift,
As the incoming waves per second get a boost,
Or a blue shift if it's moving towards us,
The incoming waves per second getting reduced.

He found that Andromeda was heading towards us,
While very many of the others were receding,
And some very quickly, at that.

The nebulae, he suggested,
Are "stellar systems
Seen at great distances".

Something was blowing
Our universe apart,
And something else was holding
Each galaxy together,
Not to mention that we could
Now surmise the Big Bang,
And that most of
The universe was missing.

POLITICAL SYSTEM BAR JOKES

A Capitalist walks into a bar...

"We don't serve Capitalists here!"

"You do now; I just bought the place."

A Socialist walks in...

"We don't serve Socialists here!"

"Can't we at least share a drink!"

A Communist walks in...

"We don't serve Communists here!"

"Enjoy Siberia."

A Platonist walks in...

"We don't serve perfect forms here."

"OK, just give me some of Greg's wine."

A Pope walks in...

"We don't serve Popes here!"

"My church has a lot of problems."

"OK, no drinking, but we can talk."

A ToeQuestor walks in...

"We don't serve everything here!"

A Melanist walks in...

"Where the heck is everyone?"

A Realist walks in...

"Finally, a servable customer."

Bacon and Eggs walk in...

"Sorry, we don't serve breakfast."

A grasshopper walks in...

"We have a drink here named after you."

The grasshopper says, "Bob?"

A five-dollar bill walks in...

"Get outa here! We don't serve your type.
This is a singles bar."

Two hydrogen atoms walk in...

One says, "I think I've lost an electron."

The other says "Are you sure?"

The first says, "Yes, I'm positive."

MORE POLITICAL SYSTEM BAR JOKES

In turn, all these different guys
Walk into their political system bars,
With their cows, asking for a drink...

The answer from their political system is given.

Feudalism...
You can have Bailey's Irish Cream
Without the creme,
For we are the creme de la creme.

Pure Socialism...
You can have as much milk as you need.

Socialism...
Go have a drink at your neighbor's house.

Bureaucratic Socialism...
Your drink is out in the barn,
Along with everyone else's.
The eggnog is there, too.

Fascism...
You can buy us a drink instead.

Pure Communism...
Your milkshake is out in the yard,
With your neighbor's.

Russian Communism...
Sorry, we drank your milk.

Communism...
Get in line.

Dictatorship...
Here's shot for you—rest in many pieces.

Militarism...
You're in the army now;
Here's a powdered drink.

Pure Democracy...
Go ask your neighbors what you're allowed.

Representative Democracy...
It's not election day yet.

American Democracy...
Drinks are flowing free; all on the house.

Democracy, Democrat-style...
You can only have half a drink;
The other half went to taxes.

Democracy, Republican-style...
This bar is much too crummy for you.

Indian Democracy...
There will never be any milk to drink.

British Democracy...
The diseased milk had to be
Poured down the drain.

Anarchy...
Your neighbors took everything.

Capitalism...
The tap has been downsized.

Singaporean Democracy...
You need a permit for that.

Hong Kong Capitalism
(alias Enron Capitalism)...
We can't account for the drinks.

Totalitarianism...
Your drink never existed.

Foreign Policy, American-Style...
We gave it away; it returned as war.

Japanese Corporation...
The bottles are too crowded to get one out.

German Corporation...
The master drinks are all on vacation.

Russian Corporation...
The Mafia took the Vodka.

Italian Corporation...
Love makes the world go ‘round.

French Corporation...
All the drinks are on strike.

Surrealism...
Two giraffes playing harmonicas
Drank everything.

WHY DO WE EXIST?

It was possible.

Does it have meaning?

Only what we give it,
As we are free to do so;
There are no strings attached.

Is this scary?

No, for we have
The ultimate freedom to be,
Within our own form. It is liberating.

THE TOEQUESTORS' DRINKS

She was the Label on the wine bottle,
Ever noting the ingredients of the fine times.

Greg drank Thuderbird 2011,
A wine that was just fine before its time.

Mikal swore off of the nasty stuff,
Drinking in an experienced education.

Lloyd joined three glasses into one,
The triadic modal logic.

Graham had one too many
On St. Patrick's Day
To make up for two months dry.

Mel's glass contained nothing.

Some glasses were half full
And some were half empty;
The engineer said that
The glasses were twice as large
As they needed to be.

THE DANCE OF LIFE

We are part and parcel of everything—
We are the cosmos; we are life; we are love;
We are all that is; we are the creator
Of the dance as well as the dancer.

LOVE

Starlight stabbed the utter darkness of night,
Causing new ideas to wink in their joined mind
As sparkling thoughts from the eternal flame,
As all the while the Cosmos played rhythm
To their merged and singing souls.

The night winds began to blow,
So the lovers nestled deeper into the leaves.

"Hold me, it's getting cool,"
She said when they were under their cloaks,
Using them for blankets.

He held her snug, his front against her back,
Spooning, not forking, until they were warm.

Then she turned and kissed him.
"As long as love's kisses can live," she said,
"Neither age nor wear on our life will show."

He sighed, growing younger,
For their love was very beautiful.

"We are wealthier than
The richest Sultans," she said.

"I pity the poor Sultan
Even with his power and status
He's not as free to live as we are."

"Yes, we are poor but rich, free yet home,
Famous but unknown."

"And the poor Sultan is stuck on his throne."

"And we're immersed in
The boundless stream of our love,
Whereas the Sultan has only
His paid-for-love harem."

"I'm realizing you now
With my whole body, mind, heart, and soul."

"They work well together, don't they?"

"Of course, they were built together
And so they weren't meant to operate separately."

"Love is reason enough for all that we do."

"Through love, all things are possible."

...

"Let us talk of love.
Let us say what it is and glory in it,"
She requested.

"The truth of all truths is love,"
He offered.

"What is the ultimate source of love?"

"Perhaps its source springs from Heaven above?"

"I don't know,
But its rhythm resonates within us,
In depths unheard of,
Plus, the rhyme of 'above' for 'love'
Is worn out."

"Where?
And 'of' is a good rhyme for 'love'."

"Somewhere deep,
Beneath all our words and thoughts,
Somewhere in our unsounded fathomless deeps,
Even that beneath which stirs
The bonding hormones."

"What is love?"

"Love is giving—with no motive toward
Getting anything back in return.
There's not even a hint of taking
Involved in giving love, because, for sure,
Taking is the opposite of giving."

"Of course; I will graciously receive
Whatever is given to me,
But I will never take it.
I will never ask for it.
I will never demand.
I will never enclose you in a cage.

"In fact, I will enhance you
So you can give even better love
To all those of the world."

"Let us give kindness to everyone in turn."

"Yes, because if you keep your love,
You will have nothing."

"And if you give your love,
You will have everything!"

"Love is more than
Just words of sentiment—
Love is action."

"Yes, one small and lovely action
Weighs much more on the scale
Than an infinite number of sentiments!"

"Sharing and caring are the reasons for giving."

"Love grows for friends and lovers
When they let it flow freely,
Beyond any confines.
One wants their partner to
Be fulfilled in every way,
Even if those pursuits take
That partner away for awhile."

"Unconditional love can never bind—it bonds."

...

"I give love to everyone
In whatever way is appropriate."

*"There is a lot of love which can be given.
Love never gets used up! It is boundless."*

"I, too, have found that
The capacity for love is infinite.
Arithmetic theory does not apply to love,
For when love is divided amongst the many,
It is not diminished in any way.
Sure, the time spent is diminished,
But not the love—I can still fully love!
In fact, each love seems to grow
To exceed the entire lot.
That's the paradox."

*"There's no good reason to ever withhold love.
Why consign someone to cold oblivion
By not sharing your love with them?
Of course, some must do otherwise
Out of tradition and moral method,
Or from bonding and commitment."*

"Give all the love you can give,
And then some."

*"Yes, since the sum of love's parts
Exceeds the whole,
One can keep on giving and giving love,
Never the less."*

"And, with a such many faceted life,
One improves,
And then one can give even more love
Thereafter as a more complete person."

*"Yes, life is more like a vast mosaic done
Than a focused beam of the sun.
There are many parts of the collage."*

"That's because few outside
And lengthy pleasures are lent to us.
We must therefore build
A stained-glass window of small ones."

*"Yes, every piece of the puzzle
Is just as important as every other,
For together they support each other
And make up the entire picture,
A masterpiece.
It takes a lot of pieces
To fit around all the sides of a person.
No one interest can match one on every side."*

(Love, when divided, diminishes not)
...

"A complete life sparkles like a diamond.
Each facet of the diamond
Contributes its view of the world
And adds to the lustrous effect."

*"Friends and interests are
The shimmering glints and gleams
Of reality's sparkle."*

"Each face of the diamond
Enriches the view of the other faces."

*"All of the facets reflect off each other,
Combining and then building
Into the overall brilliance of life."*

"Which makes you a more rounded person."

*"Which in turn adds to the luster
Of your individual pursuits."*

"Which therefore makes
The diamond even brighter still,
And so forth, and so forth—
It is self perpetuating, and of infinite growth."

"Love is the key to everything."

"Reason and passion merge into love
When truth, goodness, and beauty
Make their rendezvous."

*"Love is made up of truth,
Goodness, and beauty—
All three are clearly seen within."*

"They're intertwined as the eternal triad,
Woven into the perfect romantic braid
As its weft, warp, and wave."

*"And yet they're each different aspects
Of the same ALL."*

"For example?"

*"When a deep truth is intensely known
And stripped of all its clothes,
Then what is left is beauty."*

"Beauty is the reality of truth's meaning.
Would this be the name of the rose?"

*"I don't know, but beauty blooms,
As it were, like a rose from the soil of truth."*

"To know beauty, one must also know sorrow,
For if you're alive enough to experience beauty,
Then you're also vulnerable enough
To be exposed to its opposite twin of melancholy."

*"If we lived as figures in a painting,
Then we would never have to
Face death or sadness."*

"That may not be so great as it seems,
For what is deathless is also lifeless."

"Once I had a beautiful love with a person.
It was painful when it ended.
My reason's light began to depart.
Darkness was rising in me,
Beginning to snuff out my spark."

"What did you do?"

"Well, I gave the feelings their due.
I duly visited the shrine of sorrow.
There I found, inseparable from truth,
The beauty that had given rise to my sadness.

Upon realizing that,
Rhythms soon rose from the depths of sorrow.
I began to sing and celebrate the very song
Whose sweetness had broken my heart."

"So, the haze couldn't stop
The brightness that it veiled?"

"No, it couldn't, even though a dark fog
Had sunk and swelled all through me."

"Your love, beauty, and joy
Flowed like rays of sunshine?"

"Yes, and burned the mist
Until warmth prevailed."

"You're a positive thinker."

"It showed where my love and caring had gone."

(Love = Love / Infinity)

GOD

When you get to the eternal fundamental,
Like God or Tao,
Everything was there at the beginning.

But your theory says that Life is inexplicable
Without a LIFE behind it,
Wo why so instantly
Go against your own theory
And stop there for Something
With a great deal more Life to it, like God,
Even having a humongous size Life,
When you require it for the much lesser case?

Who is the law giver that gave nature its laws?

Everything was already there.

(How about the middle ground
Of a Godless TAO?)

The Eternal Ground-State How,
Being of forever and now,
Is the Way of the TAO
To which we all might bow,
For it is/was one big WOW.

THE RARE BOOK

The lone jewel encrusted 'Great Omar',
Now worth over 20 million dollars.
Sunk, with the mighty Titanic.
I plucked it up from the North Atlantic.

AT THE BOTTOM OF IT ALL

The last watch fire, that of mathematics,
Lights the shadows of the universe,
Telling us much about its machinery;

And, yet, there is a kind of mysticism
About this and its Platonic forms
And ideals of perfection;
So, although no one
Has been killed in its name,
It requires a kind of faith
In what magic lies beneath it;

But, perhaps, what is really there
Beneath and at the bottom of all,
Are statistics and probabilities
Averaged over large numbers of small events,
Which, though math-like come to be,
Are not exactly the root mathematical formulas;

So, perhaps math is not at the bottom of all
Although it is very much amenable
To the emergent and secondary patterns
That we observe and measure thereafter,
Being very effective in describing that “real” world.

It’s just that, as Lee Smolin sort of said once,
About Platonic forms being underlying,
“A flower is not a Dodecahedron”.

Is the universe, and even more so the world
A reflection of some perfect mathematical form?

Or does the world rest on the kind
Of statistical methodologies
That underlie our understanding of biology?

Physicists, unlike biologists,
Wrestle not with reality but
With mathematical representations of it.

This is a great and masterly art,
As is that of a painting artist,
The high beauty obtained
Not from reproducing nature,
But from representing it,
With the addition that
A physicist's greatest creations
May even truly capture some of
The deep and permanent reality
Behind mere transient experience.

There can be moments of blissful clarity,
A rare combination, indeed,
Such as when one
Really comprehends Newton's laws,
And realizes simultaneously
That what one has grasped mentally
Is a logic that is realized in each of
The countless things that move in the world.

And, yet, neither Newton's nor Euclid's laws
Completely capture the world,
But are still a fine mirror of it,
Although not the finest and
Truest mirror of reality;
Plus, there are areas that
Can't be completely captured by math.

And, thus, what is both wonderful and terrifying
Is that there is absolutely no reason
That nature at its very deepest level
Must have anything to do with math directly.

In many cases, there is a simple,
Non-mathematical reason
That an aspect of the world
Follows a mathematical law
On a subsequent plane.

Some systems have an
Enormous number of parts,

Such as why the air is
Spread uniformly in a room,
No mystery or symmetry being required,
Or how the force on a rubber band
Increases proportionally
To the distance stretched,
This reflecting nothing deep,
As the rubber band force we feel
Is a sum of an enormous number
Of small forces between the atoms.
Each of which may act in a complicated,
Even unpredictable way,
To the stretching.

A Platonist nightmare, then, would be
That, in the end, at the bottom,
All of our laws will be like this,
All the regularities turning out
To be more statistics,
Beyond which lies
Only randomness or irrationality.

It must always come to this,
As we already see in biology:
That the tremendous beauty
Of the living world is but, in the end,
Merely a matter of randomness,
Statistics, and frozen accidents—
For which the capture of there can be
No one, single, and beautiful equation.

(Gleaned from Lee Smolin)

The Equation
Of Eternity:
0

THE EMERGENCY

I was standing on the shore
Of the Hudson river,
North of the Bottini Gas and Oil Depot.

Rivulets of oil were surfacing everywhere
And they soon began to catch fire.

The tanks burst into flame
And so the town was being evacuated
Into the overflowing train station.

I floated and swam across the river,
Paddling on a pad of some sort.

The situation was nearly the same
On the other side.

I entered their train station,
Which ws even worse,
For every track had trains on it.

I walked between the trains
Where their was less and less space,
Finally crawling onto the face of an engine;

However, as it began to move,
I could tell that it was heading underwater,
So, I managed to maneuver around to the side,
Crawling up through a very confined stairway
That was very hard to manage,
It twisting and turning,
The water level rising right with me.

The train dove under the water,
Surfacing a little while later amid
A flotilla of very surprised refugees.

Well, it was only a dream,
But who or what scripted it?

IN CONCERT

The orchestra assembles
In the infant,
As the players arrive,
Section by section.

The separate agencies
Of which the baby is composed
Have to settle into place
And do their tuning up.

Nerves need tightening and balancing,
Pipes clearing, airways opening,
As well as whole ranges of tricks
And minor routines needing
To be practiced and be made right.

These subsystems that will someday
Compose a system have as yet
Hardly begun to acknowledge one another,
Let alone to work together
For one common purpose.

These parts will perform one day
In the concert called life's *Magnificat.*

Before this composition,
The infant has many selves,
For there is not yet one experiencer.

The conductor now arrives onstage,
But he only plays a minor role,
Providing some reference points
That assist with the timing
And punctuation of the playing.

He does not bind them
Into one organic unit,
For the flow between the players
Is something else—
The very act of making music

In which they act together
To create a single work of art,
Participating in the common project.

In human and nature life alike
These parts will only belong together
Just in so far as they are
Involved in the common purpose
Of creating that instance of life.

This unification that arises
Is not through the conducting power
Of some supervisory Self
Who emerges from nowhere,
But arrives through
The power inherent
In all the sub-selves, via
Their own self-organizaion,
For it is the nature of
These players to play.

The infant's symphony orchestra tunes up,
Experimenting with half-formed melodies,
To hear how they sound for themselves,
And, remarkably,
To find and recreate their sound
In the larger group sound
That is beginning to arise all around.

See, now,
How several little alliances are forming,
The strings are coming into register,
And the same is happening
With the oboes and the clarinets.

See, now, further,
How they are joining together
Across different sections,
How larger structures are emerging.

You are the dancer and the dance,
Your movements now being shaped

By the sounds of the instruments,
Your body absorbing
And translating everything heard.

One cannot make just one dance
For the many different tunes,
For that would be graceless and chaotic.

See how each of the instrument players
Is watching you, the dancer—
Looking to find how,
Within the chaos
Of these body movements,
The dancer is dancing
To the player's own tune,
Each player wanting the player
To be its own,
To have the dancer
Give form to its tune;

And, yet, to achieve this,
Each must take account
Of all the other influences
To which the dancer is responding—
How each must accommodate to
And join in harmony with the entire group.

Now there is but one orchestra
In one body of the one dancer,
Making a single work of art.

ME

Austin, taps into the external river of thought
To gain scientific insight into all life's mysteries,
Plus gaining answers to his bar exam
For becoming an attorney at law.
(But wouldn't that be cheating?)

(A pattorney)

THE GLOBAL WARMING DATA IS IN

It is no longer "if" but "when",
"Where", and how fast.

The impacts are already obvious
In the extreme north:
The Arctic ice melts dramatically,
Polar bears drown,
And Inuit hunters are forlorn.

What will be the global pattern
Of the coming human resettlement?

It will be in the northern high latitudes—
The "Northern Rim" of the U.S.,
Canada, Russia, Denmark, Iceland,
Sweden, Norway, and Finland,
Where -40 degrees will no longer
Be an impediment,
Not that it will be easy,
But, it is also the greatest untouched
Mineral, water, and energy reserve.

Russia's Siberia is nowhere near ready,
As its inhabitants have come and gone,
With no real planned infrastructure
To begin with.

Alaska and Canada are ready,
The U.S. army of WW II having
Injected massive infrastructure
Into those regions where
Human settlement
And economic activity continue today.

It will be no utopia, though,
For it will be but a conversion
From land that is hardly livable
To land that is somewhat livable.

MAKING UP GOD/BRAHMAN, EVEN FOR HAITI

And then making up further
What He does, and why:

"God is Love."

Fine, to a point, meaning that Someone
Has a lot of bonding hormones,
But, being a being or Being
Takes a village of constituents;
God is then a kind of alien,
But is not first and fundamental
And so is not 'God'.

"God is outside of all this
Since He is an immaterial spirit."

Made-up.
Can't honestly be preached or taught as fact.

So, when one makes something up,
Purely and solely considering
The felt-sensation state of being
(And further that it may have been wired
Due to indoctrination and enculturing),
Contradictions always come along to haunt
And expose the misrepresentations
That imagination came up with.

Then, similarly, the theory tries
To twist and turn to adapt somehow.
If it can't, it is merely repeated,
For it has become a well grooved
Brain-wired response.

It's not enough that God is made up,
But then people even go on to speak for the Guy,
Making up that part as well.

He is everywhere and intervenes in everything,
They say, and so it is He that causes earthquakes.

This doesn't sound so great,
Especially in light of Haiti's devastation,
So, then, they make up more to say that,
Yes, He does cause natural disasters (acts of God),
But it is to teach a lesson about life.
Yet, earthquakes only occur near plate faults,
And Haiti is also in the direct track
Of many hurricanes (4 hit there in 2008).

It also floods very much from rain
Since it has been deforested;
So, it is not cursed or being provided
As an example of suffering to learn from.

To get around this,
People then make up more nonsense
Such as the Haitians of 200 years ago
Made a pact with the Devil
(Perhaps because they used voodoo).

Even some of the God-believers
May laugh at the concept of a Devil
(What a weird belief!),
But aren't there supposed angels,
Good and bad,
And other good and evil spirits
That are really no laughing matter to believers?

Still, the original Haitians who made the "pact"
Are no longer alive, nor has Haiti seen
Any great success from any "pact";
They are now also a 96% christian nation,
One that should be in God's favor.

Made up imaginings are arbitrary,
And so they have no real bearing
On life on Earth, and, being unprovable,
Have NO REAL BEARING AT ALL
Since there is then no connection.

THE KNOWN

When one remains in the known,
One can state truths, such as
"There are vehicles with motors on wheels
That can take you places,
Or that water is H_2O, etc."

When one ventures beyond all of the known facts
One can only [honestly] say that
"What I am teaching you is unprovable".

The audience may then fade away,
But that's always the disclaimer
That should be supplied.

Sticking with the knowns of existence can't lose,
Which is why existence always trumps essence.
So, in this realm, there is always freedom to be
And to make one's own meaning in life,
For the unprovable can never be a factor.

In other words, "we can't know", for example,
About the nature of the Original Essence
Beyond the ground-state, if one is proposed to be,
Still results in the same honest stance
As when one 'decides' that any 'beyond'
Is not necessary to our
Ongoing human existence
In the here and now.

Humans being humans though,
May wish to mislead for the sake of their cause,
Always saying 'is' instead of even 'seems';
Yet, even 'seems' would be inappropriate
For the unknown unprovable.

For example:
"You will (wrong usage) incarnate
And come back (wrong usage)
As a higher or lower being (wrong usage),
Perhaps even eventually

Ascending to sit with Krishna
On planet X (many wrong usages)."

Imagine religious preachers
Being completely honest,
Above and beyond any of them
Who were indoctrinated to believe.
That we are free to make
Any meaning out of life for ourselves,
Even beyond the known is always true,
For there is no proven meaning 'beyond';
It's in the promotion of it to others
That one must be sure to state
That the part beyond the known
Is totally and absolutely unprovable.

Stick to the knowns of lived existence
And one will always have something;
Venturing beyond is fine, too,
But one will then only 'have'
Something arbitrary as a 'guide'.

"God-Man", along with "God",
His father, and "His creating the universe",
What He does, what He wants,
And what He will do for us, etc.,
Are all beyond the known—
As the unknown unprovable—
And so they remain arbitrary
In what they 'say'.

The whole argument of
An Intelligent Designer,
Some supernatural agent
Outside the laws of physics
Would seem to be a category error.

We have evolved to "see"
Agency, generally, in nature,
But science teaches us that agency
Is an emergent property,
Not a fundamental property of reality.

Stay with the known,
For there is no empirical evidence whatsoever
For agency which transcends life processes;
In fact there is a whole mountain
Of empirical evidence
Which reveals a lack of agency
In nature without life.

That, together with the
Problem of infinite regress
Kills the intelligent design argument
Dead as a doornail.
...
And there is some icing on the cake:
If life is designed the Designer
It seems less intelligent than the designed
Because of the rube goldberg nature
Of biochemical and biophysical processes.

Again from what is known from living (what else?),
Ultimately the end of the chain of explanations
Must be simpler than the thing explained,
(Lower Kolmogorov complexity
And minimum message length)
Which is why a Supreme Designer
Cannot satisfactorily serve as
The first link in the explanatory chain.

However, even that not withstanding,
All that of 'beyond' is still of no concern
Since the unknown unprovable
Is not in our realm (can't know),
Thus still making us free to be.

Any imagined 'guide'
Would have to be [honestly]
Said to be quite arbitrary,
And we do see many arbitrary
Interpretations in the world,
Whose proponents are usually
'Atheist' to all but their own imaginings;
I am just atheist to one more than they: namely, all.

THE MEASURE OF ALL THINGS

The right brain can entertain
Ungrounded notions
That are unfettered by detail,
But then, eventually,
Come the details that
Ever haunt the notion.

And this is even after the mistake
Of being human-centric,
The stance that man
Is the measure of all things.

It is neither meek nor humble
To suppose that humans
Are special and/or deserve
Reward in an afterlife.

The time of our universe happening
Is not of any special moment;
Therefore, any universes could be,
Any time or even at the same time.

Ours is not even the center of all.
There is no Earth-centric, human-centric,
Time-centric, or even universe-centric,
But only humility.
It is pride that wants one
To be King or Queen of reality.

Yet, the right brain wants what it wants;
Still, energy/material
Is not marked holy or unholy;
It just is [neutral].

THE RUBÁIYÁT PUBLISHER'S GEM

These pearls of thought in Persian gulfs were bred,
Each softly lucent as a rounded moon;
The diver Omar picked them from their bed,
Fitzgerald strung them on an English thread.

LOVE AND SCIENCE

I am indeed a quite romantic, but not incurable.

I do science, too, for if one didn't,
They wouldn't even know of bonding hormones
And other chemicals beneath.

All they would then have to go on
Would be the correlated higher state of being,
This being quite a handicap for analysis,
For then they might dream up
All sorts of ungrounded notions
That are not based on
The correlated state beneath,
From which the higher state of being is shielded.

We were talking of love being
Based on bonding hormones,
Within the larger context
That all of our state of being
Sits stop a physical basis
Of electronics and chemicals.

This knowledge is the universal acid
That eats through any and all
Superstitions and folk tales!

The bonding hormones, of course,
Then involve more of our chemicals,
Through the central nervous system, and the brain,
It emitting those feel-good opiate endorphins,
Which are more chemicals, and so on.

Chemicals R You.

Our state of being is perfect for everyday life,
For that's where we live and operate.
For the full analysis of the state of being
Such as we do on ToeQuest,
We must consider what's beneath,
For without this, imagination can go wild.

Conclusions based only on the feelings,
Thoughts, and sensations at the state-of-being level
Will not reveal the shielded essence beneath.

When one has sensations and feelings of other realms,
Such as in meditation, visions,
And imaginations of imagined ultimate realms...
One has not even left the state of being itself.

There is a subconscious state beneath,
Of electronics and chemicals,
The neural states of correspondences
To the state of being,
And, for what it's worth (a jumble of nerve impulses),
This state is not even reached;
However, science informs us about it.

QUANTUM COMPUTATIONAL UNIVERSE(S)

By traversing all possible initial arrangements,
This Infinity/Possibility/Potential
Came upon our particular solution,
Perhaps among many other good ones.

Whether all universes exist in the actual
Rather than just some promising ones
Becoming from the possible
Remains to be worked out.

Did mammal consciousness
Bring ours into being somehow,
For that is like the universe knowing itself?

THE PROOF OF THE REAL

An 'absence of something' is impossible.

1.
There is 'something'
That one's consciousness deals with.

2.
If the 'something'
Is a projected 'illusion',
Then the projector must be real,
And so that is something.

3.
A total lack of anything (Nothing),
While seemingly possible,
Was not the case,
For there is something,
So, again, the absence-of-something
Is impossible.

(Done, but for some questions
And the manner in which some basics
Can go all the way up to account for our being.)

4.
What is the something, precisely?

We can't know, precisely,
Since there can be no "precisely" there;
So, down at the simple
And uncaused ground-state,
There is not much to know.

5.
What does 'something' seem like to us?

A bunch of stuff moving around—
A movement of appearances actually,
That were emitted from the ground-state.

6.

Do some humans make much ado
About the stuff moving around,
Especially its basics beneath?

Yes, they make up that it is 'God',
This now being their last resort;

But, the composite complexity of being
Cannot reside there,
As the First and Fundamental,
Nor could any being;
But only above and beyond,
As its parts would be more fundamental;

So, 'God' cannot be the projector/designer,
Although smart aliens could be,
If they could manipulate
A heck of a lot of data.

LOVE ID

Identity is not lost in love,
For true lovers do not sit looking
Only into each other's heart,
But, rather, look outward,
Each in the same direction.

It is a seeming violation of arithmetic
That in love two become much greater
Than one plus one;
And that the two, nevertheless,
Do not become one,
But remain as two,
Yet still share the same vibration
In their souls.

THE SIMPLE BASIS OF BEING

As for forces, which is just a prelude here,
We note that two of them are transitional,
The Electric and the Magnetic,
Each giving rise to the other;

And that two others are oppositional,
The Weak and the Strong,
The Weak promoting changeability,
The Strong promoting stability.

Gravity is then left as a blend of all.

(Strong vs. Weak) [Gravity] (Electro <—> Magnetic)

So, would oppositional and transitional pairs
Work for our human being as well?

For us humans, all is of the
Movement of *Appearances,*
These *Movements* giving rise
To notions of time...

(Past into Future,
Or the Then to When through the Now),
Transitional in only one direction;

While *Appearances* beget notions of
Matter lumps, in a place of Space...

(Matter and Space, or the What and Where),
A kind of an opposition in that
The knots of matter are separate
From the gaps of space in between;

Or, in short, all seems to be the
Movement through time/distance
Of Matter in Space.

(Matter vs. Space) [Being] (Past —> Future)

We will see that our being is composed
From these simple notions begun,

For *movement* grants time—
The Then and the When
Of the Past and the Future,
Via change;

While Matter is the What,
And Space is the Where,
Via 'clumps'.

The blend of all these would be
A kind of spirit of life.

These fields then further combine:
The What/Matter + When/Future field
Becoming the Progression
Of matter into the future,

And the What/Matter + Then/Past field
Being the History
Of the matter past—what has occurred.

The When/Future of Where/Space field
Makes for Wishes, hopes and dreams
In the future place of space;

While the Then/Past + Where/Space field—
Begets Remembrance of memories
In the past space.

The emergent fields then further combine:
Learning becomes of Remembrance and History;
A Change of Outlook becomes of Remembrance and Wishes;
A Change in Structure is Progress from History;
And *Vision* is of Wishes and Progress.

Then at the next higher stage,
Being **Creative** is brought forth
From *Learning* combined with a *Change in Structure;*
Direction results from *Learning* and a *Change of Outlook;*
Growth is the *Vision* for a *Change of Outlook;*
Planning is the *Vision* for a *Change in Structure.*

Finally, **Creating**, **Direction**, **Growth**, and **Planning**
Compose one's being—The Who.

(summary):

~~Nothing~~ (Why) —> Possibility (How)

!
V

{ [Space(Where) <— Appearances —> Matter(What)]

+

[Past(was Then) —> Movement —> Future(will be When)] }

!
V

... Fields further and further combine ...

!
V

Being
(The Who)

DIA CONTINUES

Questor and Passiona recently
Addressed the new recruits:

No matter what agencies we use or blend into,
We are the Ninja Empire and can act as that alone,
Going where governments and even CIAs cannot go,
As we are not a nation, nor represent any in particular;
This is why our GrandMaster is higher
Than the leaders of all countries.

We do good only for the sake of good,
Performing the task and then getting out,
Not even sticking around to enjoy the scenery;
Your reward is the satisfaction of a job well done,
Along with shelter, food and travel.

We have chosen many from the martial arts,
As they have great discipline of body and mind.

We do not kill innocent bystanders
If they've seen our faces;
In fact, we don't kill them at all.

We use two backup teams, not just one,
But they rarely come into play.
That is where you will start.

VITALITY AND EXUBERANCE

With pep, zing, zip, oomph, vim and vigor,
We bounce along with spirit and fire;
Enthused by life's spirit energy of the zest,
We know that this life is one of the best.

BRAHMAN YET AGAIN

While it is a fine wish
That some Guy named Brahman,
God, or the Designer
Is First and Fundamental
And has creativity,
Does planning, intervention,
Manufacturing, and has even
Perhaps the emotion of love,
It is just not so,
Even doubly and redoubly not so.

A beingness or Beingness
Takes a village of constituent parts
And so it cannot be First.

The parts of all things are itsy-bitsy stuff
And so are not really anything
To speak of, much less holy.

The ground-state was around forever
Since Nothing cannot produce anything;
So, there was no creation or Creator.

One is looking in the wrong direction;
Complexity occurs above
And beyond the simple.

The idea just got grooved
Into some neuron assemblies.

So, no Big Guy at the start,
Much less some ULTIMATE GUY.

So long Brahman, God, and so forth;
You were all a fine myth
Based on the human family experience
And the tendency to believe in nature spirits.

It's also not correct, nice, or ethical
To preach and teach theory as truth.

Making things up even further,
Beyond the initial making up,
As is ever done,
Always leads to more contradictions.

GRIDLESS

Living off of the grid
In the volcanically made
Island counties of Oahu and Maui

We don't get into those pot plantations,
Plus they are well hidden.

As the motorcycle races down the Pali highway
At 90 mph the ultimate symphony begins to play
(Emotions in the state of being);
Miss Adventure rides on the back.

The motorcycle is the
Generator/charger for the laptop,
Which in turn is still the phone, the mail,
The jukebox, and the TV/movie theater.

We cross the deep blue ocean
Aboard the ferry to Maui.
There is no internet in the interior
But only in the towns.

We fly the gliders on the updrafts,
Getting closer to the demigod, Maui.

— Austin Ho, Ho, Ho

(Tiger Woods says "Where, where, where?)

THE RUNAWAY MIND

ETs or Gods?

I'll vote for ETs.

The one property of the human mind
Which gives us our capacity for imagination
Is that we project our consciousness onto the world.

I suspect this might have much to do
With our development of language,
For we began to tell stories
About the natural world in anthropic terms
Sometime early in our evolution to Homo sapiens.

This permitted information necessary for survival
To be passed down through the generations.

So, ideas about spirits in the forest,
Totems, demiurges, gods, etc.,
Were in nature-religions a way of telling
About the cycles of life and seasons —
When the fish came up river,
When the buffalo migrated,
When a certain corn plant went to seed,
And so forth.

We do these things today, of course;
Consider in the USA at Superbowl time
With the use of emblem totems
For the sports teams.

The writing of fiction is another example;
The author projects their mind through a character
Onto words on a page which are then projected
Into and out of the mind of a reader.

Einstein's imagining what would happen
If he were on a frame moving
With an electromagnetic wave
Is also such a projection.

The idea of space aliens and interdimensions
Is frankly just as much a projection
As is our mental projection
Of our conscious framework out "to infinity,"
Which is this thing we call God.

The space alien,
Along with more mundane ideas of UFO,
Are similar projections of our minor forms,
Which in the past took the form
Of angels, demons, satyrs, and so forth,
But, at least ETs seem possible.

Further, as time went on
These projections assumed celestial dimensions,
Where the heavenly hosts of the Bible
Are vague ideas about angels identified with stars.

In our modern world
These ideas have assumed an updated
Or scientific form of the ET or space alien.

Our capacity to project our emotional basis
Onto them is many and various.
Some aliens are friendly and benevolent,
Like Spielberg's "ET"
And "Close Encounters of the 3rd Kind",
Or Sagan's "Contact."

Other aliens are less than friendly,
As in the HG Wells Martians in book/movie forms,
The "Predator" movies, and so forth
Unto the hideous and almost implacable ones.

In the end we are projecting
Our mental and emotional framework
Onto the exterior world
With these ideations and fixations.

So do ETs exist? Probably.
I will say that I think the universe
In the FLRW setting is $k = 0$,

Which makes space infinite.
(A simply connected, homogeneous,
Isotropic expanding or contracting universe)

The field theory content we observe
Does not saturate the Bekenstein bound
Until about 10^{26} light years out,
And beyond that might be
Other "pocket universes"
Beyond an inflaton or scalar field
Induced barrier of sorts.

So, anything which is not forbidden
Is ultimately mandatory;
Thus, I see no reason to presume
That ETs are impossible,
So, I suspect they very likely do exist.

Of course maybe the closest ET planet
Is 100 million light years out.

God's nonexistence is another matter
One already best described elsewhere.

Our ideas about ETs
Tend to be reflections of ourselves;
This was carried to impossibly ridiculous forms
On Star Trek where you have aliens and humans
Bearing offspring,
Such as Spock is half Human and half Vulcan.

Nope, that is not going to happen.

Our ideas about ETs most often
Are exaggerated forms of ourselves,
Such as the bug-eyed bulbous headed grey aliens
Portrayed so commonly.

This would extend to the idea that any ET
Would project their selves as we do.

The internal mental reality of an ET

Might be so radically different from anything
We experience subjectively
That we couldn't possibly ever understand it.

This would likely be the case,
Even if we can decode
Their electromagnetic signals
And figure out how they do
Mathematics and so forth.

The converse might likely hold as well.

So the concept of a God,
Such as found in their projecting
Their internal mental reality out to infinity,
Might simply be outside
Their capacity to internally experience.

If aliens did decode a signal from us
And collect our ability to project our minds,
They might find this to be a unique way,
For what to them is an ET, to experience
External and internal existence.

Further, they might find the idea we have
Of God to be utterly beyond their ability
To internalize or understand.

Of course this point is
Likely ever more the case
When it comes to the
Particular theological ideas we have.

THE SPECTRAL

Yes, down with all those figures
Darting about in a purplish fog,
Those ghostly phantoms that are wraithlike,
Shadowy, incorporeal, insubstantial,
Disembodied, unearthly, otherworldly,
And downright spooky.

FUNCTION FORMERLY KNOWN AS FUNCTIONING INTO FORM...

I can see where God could be ruled out
As intending the first function
That went on as intended design
To it's completing form, since,
By the same rules of form needing function,
God's form would would have to had derived
From a preceding function;

But, getting back to reality,
It would seem that the uncaused
And eternal fundamental substance
Would have nothing prior to it
To intend its function;
But, then, of course,
From then on there would be cause
For its secondary recombinations
And complexities operating on upward
To form composites.

Strictly speaking,
It seems that there is no 'random' anywhere,
In this upward and onward,
As macro movements have causes beneath,
At least secondarily above
The initial uncaused level of FS;
Although we have to wonder if some micro effects
Like radioactive decay are 'random',
Since the 'beeps' seem to have no pattern.

So, only a part of the question is about 'random',
As weather and environment for evolution
Will always come about sooner or later
In some average mix of happenings
To affect natural selection;

The more important part being about
Form being intended by some base Function.
I just don't see any Fundamental
And Intentional Being doing it,

Just that a lot of tiny FS moving around
Can amount to something,
Given that it was not inert
(And possibly even that it couldn't be).

The movement of appearances would seem to provide
For a kind of 'spirit of life' for animated beings
That in themselves even reflect that concept.

THE PATTERN

Some mention that order
Requires Order behind it,
That life requires Life behind it...

This is the subjective mind
Halting prematurely,
For then all the more
Would this Order/Mind
Need ORDER/MIND behind It.

Then some time goes by
In the subjective minds,
They perhaps being immune to the objective
Or just mired in the same groove—
And then they again restate
The same old theme again...

...Such as cause needs Cause,
Again implying that,
To continue the chain,
Cause obviously
Requires CAUSE behind it,
And then I suppose more **CAUSE**...

But cause can't go on forever...

YOUNG AGAIN

I am home, back where I began.

If, by our late middle age
We begin to really live,
Although by then it's almost too late,
Then it's because our prior life
Was but a preparation:

In our forties there may have
Been more work than play
As we solidified our careers
And guided our children on;

Our twenties had demanded of us
The unsettling stresses of graduating college,
Finding a job, wooing a mate, and buying a home;

In our teens, although our hormones
Were flowing wildly,
We were often thwarted
By the cell walls of study,
Curfew, and sexual responsibility;

Only as children were we almost free,
But even then
The shadow of authority everywhere
Passed as a dark cloud.

Therefore, it is only when
We spread into middle age,
Say at age fifty or so,
That we finally reap real interest
From the dues we've paid.

We are free to live and write,
To fully create art, life, and love—
Albeit, though, that death's faint knockings
Have already sounded in our hearts,
And that time's corruption is seen in the wrinkled skin
That we may fondly try to stretch baby smooth.

A step or two is lost in tennis
And age is noted in the graying of the flower,
Although the root may still be green.

Yet, for all this,
There is a new exuberance
That never was,
A realization, at last,
Of the full worthiness of life
And of its precious pleasures,
Of the promotion of one's spirit
To a higher plane—

And the complete removal
Of oneself from parts of life
That suddenly appear quite needless,
And a determination to live even more,
The way we would have if we could have
Ditched out of all work and worry.

Yes, the unseen but still sensed specter
Of old age still looms;
But, it is well around the corner—
Not even an enemy,
But a most inspiring presence
Which promotes living, not dying.

So, one is reborn.
This and that home improvement
Seems no more to matter so much
As does creation, friends,
Health, adventure, and loving.

EVERYTHING

Everything possible must exist.
Or, if you prefer:
What never exists is not possible.

Faith is any contrary assumption.

SELECTING FOR FITNESS (IN ADDITION TO THE UNFIT)

Natural selection can and does select, as well,
For mutations that increase overall fitness,
That is, these mutations increase the frequency
Of that allele in the next generations;
For example, a mutation
May elongate the beak of a bird
And make it thinner;
This would enable the possessors of this allele
To eat insects living in crevices in the rocks.

Or, a mutation may improve
The digestibility of proteins
And thus increase the food supply;
A fungus may acquire
An enzyme that digests cellulose,
Which few species can do—
A whole mountain of food is now available.

There may be other mechanisms
In addition to strict natural selection.
A possibility is the case of mutations
That accidentally carry both a selected property
And another quite different unselected property.
This gene may now hang around a long time,
And may by chance find itself in a new environment
Where the second property becomes of selectable value.

Jack Beans contain an enzyme called urease,
Which very potently hydrolyses urea.
To the best of my knowledge
Urea does not occur in plants!
A second example is a
Boring enzyme of metabolism
Which moonlights as
A DNA binding protein
Of very high specificity.

With respect to selecting for fitness,
One of the great all time experiments

Was done by Richard Lenski which examined
A species of bacteria in an environment
That contained a specific type of fuel
And some other substance, perhaps citrate.

The bacteria were stained a certain way
So that researchers could determine species.

When a mutation occurred,
These mutant bacteria were then placed
Into their own isolated environment
With the same fuel and solution,
And some of the previous species
Were "fossilized" via freezing
For comparison later.

As you can imagine,
Bacteria that could specialize
By living in glucose (the fuel) proliferated better,
Just as NS would suggest;
Fitness increased quickly up until
About 20,000 generations,
Whereby these mutated bacteria
Had grown about 70% more quickly
Than the initial ancestor strain,
And then growth tapered off
In an asymptotic way.

A few interesting developments to note—
Some strains developed a mutation
That negatively affected their ability to repair DNA,
Which increased the rate of mutations in those strains.

By way of numbers, some 100's of millions
Of mutations are believed to have occurred
In the first 20,000 generations,
But only 10-20 beneficial mutations
Gained fixation in the main population.

The most important mutation led to the ability
Of a strain to use citrate as a fuel,
And so those mutant populations experienced

An additional surge,
Breaking through the
Previous asymptotic growth.

This mutation depended on another mutation
That had up to this point been neutral
(Non-adaptive at the time),
But, coupled with this additional mutation,
Became potent in the citrate/glucose environment,
Point being mutations can increase overall fitness,
And can be quasi-cumulative in the scenario
Where one mutation is neutral by itself
But incredibly potent along with another mutation.

There are other factors in evolution
Besides 'regular' Natural Selection,
Such as cataclysmic events
Like asteroid impacts,
Natural disasters,
And other broad extinction events.

Perhaps had there been no asteroid impact
65 MM years ago mammals would never
Have had the chance to dominate to this point.

Gould and Lewontin were big believers
In punctuated equilibrium,
But that is still ultimately
Explained by Natural Selection.

There are occasions when
True random behavior is evident,
Such as in mutations;
But equally, the response to mutations
Is not so random at all.

An organism will find that any specific,
Randomly chosen mutation
Will fall into only one of three categories:

1) It has no selective advantage
Under current conditions—it is neutral,

2) the mutation is positively advantageous
For that species in the current conditions,

Or 3) the mutation is clearly disadvantageous
For that species in current conditions.

The vast majority of natural mutations
Fall into the third category;
The second category is the least frequent.
The frequencies of these processes are amenable
To simple mathematics, and so they are quantifiable.

As we know,
Contemporary biological evolutionary studies
Have a large mathematical component;
So, I would say that there are at least general rules
Which describe the process of biological evolution.

The idea is that mutations create novelty,
Most of which is maladaptive,
But that natural selection prefers some to others
By not killing the adaptive ones
As rapidly as the maladaptive and neutral ones.

In terms of art, a sculptor CREATES a statue
By carving away clay that isn't appropriate
To what he is creating.

Dawkins invoked the idea of a blind watchmaker;
We might consider a blind sculptor.
The block of marble is chipped away
And there is a form generated,
While at the same time detritus
Has to be swept from the floor.

I think natural selection might be considered
A creative force in the same way
A river carves out a canyon,
Tectonic activity bends and shapes rock,
Or wind and water may sculpt formations of rock.
These too are blind forces as well.

ON THE ISLAND

The dusk deepens, night's pot of tea steepens;
Silence descends, as when a gift opens;
Eventide rises. On high, Orion camps.
Our eyes catch stars like fireflies in lamps.

The ferry was continuing on to Kauai,
A rare destination,
So we remained aboard.

It's a quiet island,
One good for honeymooners,
And sightseeing,
Containing the legendary Bali Hai
And the Waimea Canyon,
Lined by a road that is
Not much of a road at all.

Amid the endless sugar cane fields
We came upon yet another tin shack,
But this time we stopped
And gave them some goodies,
And talked and stayed
Into the night
With this Filipino couple.

In the quiet of the night
We could hear the waterfalls
Rushing, way off in the distance.

They knew the names of the stars
And that, therein them,
Hydrogen was being converted to Helium
And that that was why the stars shine!

They were surprised
That I knew some Tagolog;
However, I'd spent a lot of time
In the Philippines.

...

Space gives me room to realize
That the Earth couldn't
Be much farther out in space, alone,
It rolling along a spiral arm, unknown.

The stars of space beckon,
Warm and welcome,
Being the fires of home—
Those ancient lights
Piercing the depths of time.

Look at the stars
In the depths of the night;
Hold the flames in your mind,
Keeping them bright.
Their power flows,
Energizing you from
The eternal charger—
You see the light!

The stars are my mind,
Having made my mind,
And so I'm ever inspired
By a thousand ideas
Beckoning from afar.

They wink in the mind's meadows,
Like fireflies;
They stab the darkness of naught
With their light,
For the eyes can ever catch these stars,
Like fireflies, in a jar, to make the lamp
That burns the night away.

They are eternity's running-lights;
They're the gleam in my eyes;
My smile's light
Is that of a distant sun
From long ago.

From Heaven's stars
Came my dust eterne,

For I was born of stardust
And then nourished
By the sunlight
That filled my living cup
With so many wonders of delight;
For Time's seas
Nurtured me and thee in turn.

From time, death, and dust
I thus became,
And so by this, thus, and that
I must return.

Star light is the origin of our being,
Being the source of our matter, energy—
Everything; It's our radiant spirit,
Our self-winding mainspring.

Soul to soul, it said to me,
"I'm the light, thy spirit's sight,
A beauty bold and bright,
An inspiration come from darkest night;
You're a newborn star aglow with insight."

Oh thee, of thine,
Whence came this life of mine?
I wish thee to thank for this living wine.
Oh, Nature, Father Time, Guiding Star—
Thanks for throwing me this earthly lifeline.

Our shadows are touching, in the same shade—
We embody, in third dimension made;
We kiss, drift, cross into each other's role;
Spirits open—rainbows meld in the soul.

(We still have much to know about star formation,
As it is that stars about 20 times larger than our sun
Would seem to have to be limited at about that point.)

COLOR TURNS

The color spectrum that we observe
From E/M wavelengths stems
From proteins in the cones of the eyes.

These three types of proteins rotate,
According to the amount of
Their associated primary color present.

Three primary colors;
Three corresponding types of proteins.

Nature is simple and efficient;
From the mixture of just these three primaries
Come the jillions of colors.

Color blindness occurs when
One or more of the protein cone types
Doesn't function.

Glimmers and glints are optical effects
That are not really out there;
A rainbow belongs only to you,
Moves with you, etc.

THE FAMOUS ELITE—
WHO SEEM TO KNOW AND DO EVERYTHING

The elite are best pick of cream,
Being the crème de la crème,
The best flowering of humanity, nonpareil,
Those "elected" to high society and the jet set,
All the beautiful people,
Beau monde, haut monde, and glitterati,
Those of aristocracy, nobility,
And the upper class.

INADMISSIBLE NON EVIDENCE

'God' is not admissible,
For it is unprovable and not known;
Even having no evidence;
Whereas, evolution is known and has evidence.

This is why an idea that stems from
Only the idealized state of being
Of how we feel or wish
Goes into battle only partly armed,
Plus, it is even that some of
The idealizations must be discarded.

As for how we seem to ourselves,
Note that our conscious state of being
Doesn't directly inform us
On the actual states of being beneath.

Some like to think "case closed"
About a wish for God being true.

I realize it's a personal thing
That may have been grooved
Into the wiring of the brain.

To inquire why,
One must then undertake
And develop some second level beliefs
About one's first level beliefs,
Neutrally if possible,
At this next higher stage of thinking,
Becoming rather a spectator,
To some degree,
Of one's first level beliefs
To observe whence and why they exist,
Beyond just noting that they are there,
And going with them.

And, even beyond this the perceived state of being
Is not of the same type of the state beneath us.

So, as for how we seem to ourselves,
And what we come up with,
Based on our state of being view,
Is not enough if not supplemented by science,
For our conscious state of being
Doesn't directly or entirely inform
Us on the actual states beneath.

NO MALFUNCTIONS

God, being perfect, has no evil failings,
But for the biblical tales of His ranting and railing.

These are but imaginations of stories old
That down through the generations rolled.

Imperfect humans can selfishly hate
And so transcribed these onto Fate.

Love is the only emotion of any worth
That would have been showered on the Earth.

STABLE ORGANISMS

Basically, evolving organisms
Are a stable platform,
Since they didn't die off;
They 'take on' chance to perhaps move on
To the next stable level;
So, evolution doesn't just proceed
By all kinds of chances right in a row
Like one trying to make
A fortune at a roulette wheel.

THE HEAVENS DO NOT SEEM TO DECLARE THE GLORY OF LIFE

It seems that we are lucky that humans evolved,
Much less aliens, although if we could, they could;
Yet, Graybeard's referenced searches don't find them.

The world and the universe appear
To be precisely the kind of world and universe
That is utterly indifferent to life,
As opposed to specially created for it.

The universe is roughly
93.539132753 billion light-years in diameter
And constantly expanding,
Meaning that virtually all of it
Is permanently inaccessible to,
And even unobservable by, humanity.

Scientists currently believe
That 96% of the universe is
Either dark matter or dark energy,
Meaning that a scant 4% of the universe
Is even conceptually accessible by us.

Of that 4%, virtually all of it
Is comprised of empty space
Some two degrees above absolute zero,
Which is (of course) instantly lethal to living beings.

So, essentially,
The universe is almost
Entirely off-limits to humanity,
And of that which is not off-limits,
Almost all of that is trying to kill us.

(Info from somewhere):
We occupy one planet orbiting our star.
It would be difficult to precisely measure
The boundaries of what constitutes our solar system,
But it includes, at minimum,
The orbit of the dwarf planet Eris,

Which spins out to approximately
100 Astronomical Units (AUs) from the Sun.

Each AU is 150 million kilometers,
So if we consider the Solar System to be a sphere —
I know it isn't, but bear with me here —
With a radius of 100 AU,
We get a volume of approximately
Fourteen million, million, million million
(1.4 x 10^26) cubic kilometers,
Or enough space for more than ten trillion
(10,000,000,000,000) Earths.

We thus occupy,
In rough, back-of-the-envelope terms,
One ten-trillionth of our Solar System.

Now, consider that our galaxy contains
At least two hundred billion stars
And accompanying solar systems
(200,000,000,000),
All of which are inaccessible to us
Unless we engage in science-fiction make-believe
And postulate some way to travel
At or above the speed of light.

Our galaxy, in turn is one of more than
A hundred billion galaxies
(100,000,000,000)
In the observable universe,
None of which are accessible even
With science-fiction make-believe —
The closest galaxy to us, Andromeda,
Is 2.5 million light-years away.

With a little quick math,
We can see that we occupy just one part in
100,000,000,000,000,000,000,
000,000,000,000,000,000,000
(1 x 10 ^-38) of the conceptual "real estate"
Of the universe, not counting
The vast empty space between each solar system.

So 99.99999999999999999999999999999999999%
Of the universe is basically off-limits to humans.

It gets worse.

On our infinitesimal speck of the universe,
Most of our planet is also inaccessible to us.
Over 70% of the Earth
Is covered in salt-water oceans
That we cannot stand on, live in,
Or breathe or drink from.

Of the remaining land,
Half of that is taken up
By uninhabitable mountains,
Glaciers, deserts, or other unlivable terrain.

On the tiny slice of land that is habitable,
We are subject to the uncontrollable whims of nature,
Such as the vicious tsunami I describe above.

Keep in mind, too,
That we are newcomers on the scene.
The Earth is 4.5 billion years old,
And for virtually all of that time period,
The Earth's climate has been inhospitable
To human life.

Meanwhile, the Sun
Is slowly expanding and growing hotter,
Such that will no longer even
Conceptually support life
Within the next billion to two billion years.

So the argument that this world
And this Universe
Is conducive to life,
Runs against the mountain of evidence
To the contrary,
Indicating that aliens would be
Few and far between.

At most, we have a few percent of this globe
For a tiny fraction of its history,
Which is in turn an infinitesimal fraction
Of the universe.

Either all the universe
Is "for" our benefit or life at large or it isn't;
You see that I am fair to consider either.

Yet, we have existence on a planet
That is constantly trying to kill us
With natural disasters, deadly viruses and bacteria,
Poisonous insects, and so on and so on.

We could run through the same exercise
With the development of life;
In fact, I think this intuitive disconnect
Explains why so some may go to such lengths
To reject contemporary evolutionary biology.

It just doesn't seem compatible
With human type life
Being a major happening
When we realize that
The crocodile and the coelocanth
Have been around
10,000 times longer than we have.

So, anyway, where are the aliens?

I do hope to meet some someday,
Although since we don't see their ships
That have come from so far away,
I doubt that they are around.

One would have to again retreat
To the magical invisible realm to accommodate them.

If so, then they are quite kind.
Although their curiosity can't seem
To get enough of us, through abduction,
They are nice enough to always put us back in our beds.

THE GRAVEYARD OF THE GODS

Where is the graveyard of dead gods?
What lingering mourner waters their mounds?

Where the caretakers and lawn keepers?
What powers do these gods have now?

There was a time when Jupiter
Was the king of the gods,
And anyone who doubted his grandeur
Was called a barbarian and an ignoramus.

But where in all the world
As one who worships Jupiter today?

And who of Huitzilopochtli?

In one year—and it is no more than
Five hundred years ago—
50,000 youths and maidens
Were slain in sacrifice to him.

Today, if he is remembered at all,
It is only by some vagrant savage
In the depths of the Mexican forest.

Huitzilopochtli, like many other gods,
Had no human father;
His mother was a virtuous widow;

He was born of an apparently innocent flirtation
That she carried out with the sun.

When he frowned
His father, the sun, stood still.

When he roared with rage,
Earthquakes engulfed whole cities.

When he thirsted he was watered
With 10,000 gallons of human blood.

But today Huitzilopochtli
Is magnificently forgotten.

Speaking of Huitzilopochtli
Recalls his brother, Tezcatilpoca,
Who was almost as powerful,
For he consumed 25,000 virgins a year.
Lead me to his tomb.
Would I weep, and hang a string of pearls?
But who knows where it is?

Or where the grave of Quitzalcoatl is?
Or Xiehtecuthli?
Or Centeotl, that sweet one?
Or Tlazolteotl, the goddess of love?

Of Mictlan? Or Xipe?
Or all the host of Tzitzimitles?
Where are their bones?

Where is the willow on which
They hung their harps?
In what forlorn and unheard-of Hell
Do they await their resurrection morn?

Who enjoys their residuary estates?
Or that of Dis, whom Caesar found
To be the chief god of the Celts?

Of that of Tarves, the bull?
Or that of Moccos, the pig?
Or that of Epona, the mare?
Or that of Mullo, the celestial jackass?

There was a time
When the Irish revered all these gods,
But today even the drunkest
Irishman laughs at them.

But they all have company in oblivion:
The Hell of dead gods is as crowded
As the Presbyterian Hell for babies.

Damona is there, and Esus,
And Drunemeton, and Silvana,
And Dervones, and Adsalluta, and Deva,
And Belisima, and Uxellimus,
And Borvo, and Grannos, and Mogons.

All were mighty gods in their day,
Worshipped by millions,
Full of demands and impositions,
Able to bind and loose—
All gods of the first class.

Men labored for generations
To build vast temples to them—
Temples with stones
As large as hay-wagons.

The business of interpreting their whims
Occupied thousands of priests,
Bishops, and archbishops.

To doubt them was to die,
Usually at the stake.
Armies took to the field
To defend them against infidels;
Villages were burned,
Women and children butchered,
Cattle were driven off.

Yet in the end they all withered and died,
And today there is none so poor
To do them reverence.

I wonder what has become of Sutekh,
Once the high god of the whole Nile Valley
And of more gods?

All were gods of the highest eminence.
Many of them are mentioned
With fear and trembling
In the Old Testament.

They ranked, five or six thousand years ago,
With Yahweh Himself;
The worst of them stood far higher than Thor.
Yet they have all gone down the chute,
And with them so many more.

Ask the rector to lend you
Any good book on comparative religion;
You will find them all listed.

They were gods of the highest dignity—
Gods of civilized peoples—
Worshipped and believed in by millions.
All were omnipotent, omniscient
And immortal.

And all are dead.

THE ETERNAL-TIME-COMPLETION-UP-TO-NOW PARADOX

There's nothing much more to reality,
Since the 'fluctuations' existed forever,
But, they were not time-forms, evidently,
Whatever this means,
Or there never would have been
An earliest 'record' of them.

The emitted particles that endured
Became the real,
Granting definition and movement,
And, with them, and of this,
Time became,
"Eternal" unto their end,
If they ever do.

NON-VERIDICAL BELIEFS
(Subjective)

Those claiming invisible things
Usually do so outright,
Often not even saying
That it is just a theory.

Even so, what one then
Suggests as what 'seems',
Which they don't, either,
Still often comes down
To the necessarily invisible.

The problem,
Beyond that about "found" invisibles,
Is more about what people actually believe
And why they do so.

There has been some research done
In the area of belief.
Robert Burton has written a book
Which does a fine job identifying
The evolutionary basis for belief
And then finding which circuits
In the brain are involved
(See "On Being Certain:
Believing You are Right
Even When You're Not").

The basic idea is that the feeling of being certain,
Or knowing something is involuntary
(i.e. mediated by subcortical structures)
And bifurcates into two basic categories:
Veridical and subjective.

Examples of the veridical category
Include things like the capital of Paraguay,
The number of atoms in a mole,
The prime minister of England;
While examples of the subjective category
Include things like God exists,

The Bible is the word of God...

...These being beliefs that are of a nature
Which we do not have available to us
The information necessary to know them
To the same extent that we know
That the President of the U.S.
As of Feb. 5 2010 is Barack Obama.
Yet at the same time the brain will believe
These subjective "facts" with the same degree
Of certainty as the veridical category.

What's going on?

We know that belief
Is strongly dependent on memory,
Which is paired very strongly with emotion.

This accounts for the involuntary nature
Of the feeling of certainty and conviction
About topics for which no answer
Is available, by pointing us to the amygdala
And other emotional centers of the brain
Which then link to reward centers
Most likely via the basal ganglia.

In English that means that these beliefs
Are being processed in the same manner
As those of a veridical nature,
And become reinforced
Via the brain's reward circuits.

"...Belief", judgments of "true"
Versus judgments of "false",
Is associated with greater signal
In the ventromedial prefrontal cortex,
An area important for self-representation,
Emotional associations, reward,
And goal-driven behavior."

-Taken from a study which was done in part by Sam Harris.

That is, the same region of the brain
Showed increased activity
When subjects judged a proposition
To be true regardless of the content
Of the proposition in question,
Suggesting that a common neurological mechanism
Is used regardless of the subject under consideration.

Of course this is a simplification
And ignores the activation of the mPFC
And parts of the temperoparietal junction,
But the emphasis here
Is on the rewarding nature of the belief.

Because children are usually indoctrinated
Very young, when their brains are more plastic,
This system of belief can establish itself
Prior to other systems
And helps shape the way
In which the child then views the world.

It then becomes difficult to displace these beliefs
As the brain becomes less plastic into adulthood,
Given the reward signals they produce
And the feeling of certainty they generate.

This leads to seeing agency
In inanimate matter (invisible agents),
Seeing faces in the clouds, etc.,
But, basically,
We are programmed by evolution
To seek out intention
And there are brain correlates
Associated with this activity.

THE AMAZING DISCOVERY OF FORBIDDEN COLORS

Are there any missing hues,
Unknown, hidden in rainbows, or not used?

It turns out, that,
Due to the way our visual system is,
We cannot see bluish yellow—
And I don't mean green,
Which can mix of those;
Nor can we see reddish green;

Yet, I have made
The amazing discovery of seeing them!

By looking at a square of blue
And one of yellow,
Each having a small cross in the center,
I was able to cross my eyes
To bring them together,
And then saw the elusive
Bluish yellow,
And in another cross,
Reddish green,
Although at first
They jumped all around
And even put holes into each other.

Meanwhile, I am in the bliss of seeing the forbidden.

WOOF!

Love's warp, weft, and wave,
Is the eternal, perfect braid,
Wound of the strands of Truth,
Goodness, and Beauty made.

THE UNKNOWN

No one can yet know the origin of the message of life
Or the exact cause of the origin of the universe.

We are many times removed,
First, by the brain,
From the actual real frequencies
(That senses take in),
And, ultimately,
By just not knowing
The source of the old frequencies,
Which were way back then, anyway,
Before the universe was 'born'.

Actually, we are more like
Eleven times and more
Removed from whatever
Originated the universe,
If it even was originated
(The basics could have been eternal),
The higher (in the chart below) becoming
Of the lower beneath it as
Either interpretation
Or more complexity or both:

(The Chart)

9. The higher brain of knowing.
8. The rest of the brain
(subconscious).
7. Cells
6. Senses
5. Molecules
4. Atoms and photons
3. Stars
2. Protons
1. Quarks?
0. Origination of the quarks/universe?
-1. Potential?

Our life evolved from a common ancestor,
But our 'life' cannot know
The "before" eleven times previous.

To claim that we can could truly be called delusion.

Now, then, how could a first and eternal source
Have been so well defined
Without ever having been defined?

VACATION

I'm on vacation at a secret hideaway
Right on the ocean,
Doing a study of materialism.

Everything seems real here,
But for some obvious cases of silicone fraud.

My room is so unassuming
That its entrance is via
An unmarked door in the stairwell.
No bad guys can find me here.
The town is filled with transients—
Visitors from all over the world;
Same with all the workers—
They come here from afar to work
During their summers off
From college or whatever.

My room even has a little hump
And a downslope just inside the door,
so even if any bad guys did get in
They would immediately fall down.
I'm off to the depths of the sea now.

THE STRING STING OPERATION

God had no friends,
A Being all alone,
Nor any earliest memory,
So, he took up His harp
To play some perfect songs.

The strings vibrated,
Playing forth the music of reality
As all the fundamental particles.

See M-Theory Movie trailer
Showing quark workings
and asymptotic freedom
Of free will:

http://www.youtube.com/watch?v=p93Gn_nrVIw

"Of what instrument have we been spanned?"
...
"Oh, sweetest song!"

All would have been fine,
But along came the Devil,
Existing because:

— *Positive and Negative Balance* —

Good and evil were wrought from wrong and right,
When, from nought, twin genii split day and night.

The Devil played not even his simple fiddle bow,
Nor even his wretched accordion,
But a set of bagpipes
That was way out of tune...

This dark symphony of discord
Distorted the anthem 'ne plus ultra',
It becoming even much worse than rap "music";
So, thus it is that all is not perfection

And, so, to us puppets
Many strings are attached.

Never-the-less, Brian Greene
Discovered by math alone
The 10*500 possible tunes
Of all the universes strung
And sung in so many dimensions.

In 'M Movie',
The movie of all movies,
"A Star is Born",
Namely, Greene Brain,
Who takes us
Into another dimension...
Of the con zone.

(These songs are playing
In a universe near you.)

I BREATHE

There was no one miracle of life
Leaping into immutable living form;
Slightly dead chemicals gradually
Became definitely alive chemicals, like RNA.

We even have evidence of ancient algae
From 3.5 billion years ago.
This was when liquid water became available.

It still took more than two billion years
For more complicated life to appear.
There was then, no literal Garden of Eden.

ACCEPTANCE

I accept almost everything,
Except maybe mosquitos,
Although they are probably
A necessary part of the food chain,
Because the world is perfect
Just the way it is
For the evolutionary stage it is at.

Of course, any one thing on Earth
Probably has a mixture
Of good and bad uses,
But, if we were to take it away,
The whole balance of the earth might collapse
(Do mosquitos have a use?)

Yes, we have crazy emotions sometimes,
Some of which are rather useless now,
Or almost so,
That are still forced upon us,
Very much uninvited,
Such as jealousy and anger,
And, too, children's emotions are not mature
And teenagers have wild hormones
And holes in their brains,
And both apparently
Bring great suffering to all,
Not to mention what the deviants
And low-lifes of the world
Do to us and each other—
But all this is just a stage
That the very primitive
And mostly infantile human race
Is floundering through.

I accept even death,
For without death I wouldn't even be here,
For my human ancestors evolved because of it
And if they would have lived forever
Then not much DNA would have ever changed.

I accept the natural world
And even the humans mentioned above
That are just as much a part of natural world
As anything, crazy as they are,
For, without the beast in us, for example,
there would have perhaps been no zest for life.

We have probably survived
Not in spite of being brutal,
But because of it, and at least it may have led
To cooperation for the hunt and for war.

I accept bacteria
And all the diseases they bring,
For they also ferment the soil
And aid digestion.

I accept bugs and worms,
Although I am still working on
Accepting mosquitos.

Worms, for example,
Aerate 400 tons of soil per day.
No worms, no life;
So, please give a worm a hug today.

The same with flood and drought,
For the universe has our well being at stake
Only in the most general sense,
But not in the specific sense;
A farmer's crops may dry up
Or get washed out, but, again,
If there is no water, then there is no life.

Well, anyway, it's warm here,
But there are no mosquitos,
So, I'm really extra happy about that.

AUSTIN INVENTS GOD

It's opposite day,
And I'm climbing Jacob's ladder;
So, here's my theory
Of what gave purpose to our universe.

In some previous arrangement
Of the Fundamental Substance
Long ago, in an incarnation
That we'd call a universe,
A very intelligent universal mind formed,
This taking almost 10**137 years,
But for a week or two.

The arrangement/evolution
Of this great mind
Had to happen,
Given all the time of eternity.

This mind was so smart
That it could rightly be called God.

As that universe was winding down,
God mixed a mighty brew
Containing the right proportions
Of matter, dark matter, and anti-matter,
As well as the necessary forces,
The sum of which would allow
Life to form in the next universe
Within a mere 13 billion years.

It wasn't that He was able to extrapolate
The effects of the simple
To the complexities of being,
This being well nigh impossible,
But that He had noted the right recipe
That had finally led to him
Through His forebears
After some near infinite number of
Universal type arrangements of FS
Had hit on just the right mixture.

So, He made our universe in six days,
The experiment working fine,
Although realizing that it, too,
Might eventually lead to its own
Universal Mind some day;

Upon which occasion
He would gain a friend of equal stature.

God is truly a great guy, and any bad things
That you hear about Him are just simply not true.

LOVE IN THE AIR

It was in '60, '61, or thereabouts, aged 13,
That we strolled through the woods
Behind Cynthia's house and found
A small clearing snug with trees all around it,
With a picnic table,
A space somehow perfectly preserved
From its older days.

We liked these enclosed spaces
As much as those of tree houses and such.

The dusk was yet viable in its embers
And a good enough moon was up
To grant the enchanted lighting.

The place spoke volumes of intimacy
From the summation of its whisperings.

Worries in those days seemed
To be less than those of now.

I went back there some months ago.
Her presence, psychologically, was still there,
As well as the place being just the same.

It is a magical island standing still
Amid the entropy ever marching on elsewhere.

BLANKS

One can begin to fill in the blanks
Of the invisible realm,
And then even complete many details;
However, this will just amount
To a lot of words,
For the realm ever remains invisible.

The words then meet contradictions,
As made-up words ever must.

At ToeQuest, we look into truth,
Noting the implications of theories.

There is no challenge to a mode of life
In which someone can gain
A measure of peace
By wishing one's self
Into a particular mindset.

I often envision a kind
Of slow-paced Victorian world,
Without the problems, of course,
In which I can leisurely stroll about
Enjoy the scenery and
Talk and adventure with everyone.

And so I have found such places
On this modern earth,
Such as a riverside restaurant
On the mighty Hudson
Where afternoons can be whiled away,
As well as in the archipelago of Hawaii
And, it seems, now, wherever I go.

So, one's choice of existence
Can and always does trump essence.

As a hobby, I also look into
The Theory of Everything.

References to “outside”
And “invisible” forces
May ever come as replies;
But, as always, if we use a word,
We must fully explain that word
In all its specifics,
And in light of ToeQuest
And the Theory of Everything,
We must also do that
On scientifically satisfying grounds.

Consider this stance:

The brains function is receiver-transmitter,
Sometimes from body to mind—
Sometimes mind to body.

The brain limits the mind
By restricting focus of attention.
The brain is an organ of “attention to life”
And an obstacle to wider awareness.

The brain as a limiting obstacle
Filters out forms of consciousness
Not necessary for biological needs
Which explains that freedom from body (NDE)
Results in more extended forms of consciousness.

First of all, it seems that the definition above
Already presumes that mind and body are separate,
Perhaps the ‘mind’ said to be immaterial—
A kind of a soul perhaps connecting to God,
This mind/soul also called consciousness.

I would say that if mind and brain
Can talk to each other,
Then they must speak the same language,
Thus making them material and physical.

These are a lot of words
Pronounced upon the invisible.

Nor would I think that a Creator
Or evolution would make a brain
That is an obstacle to itself;

Or that one would have to get
To near dying to go beyond
The brain for a brief moment.

And consider:

The dependence of consciousness on the brain
For the manner of its manifestation
Does not imply that consciousness
Depends on the brain for its existence.

Yes it does.
No brain; no consciousness; that's it.

The brain is not an organ
That generates consciousness
But rather is an instrument
Evolved to transmit consciousness.

These invisible details
That are merely supposed
Are, again, not available.

Consciousness reflects
The brain's doings and analysis,
Day in and day out.

All in all, science,
As seen all about ToeQuest,
Shows otherwise
To the invisible realms proposed,
TIme and time again.

DISCOUNTING EVOLUTION

Those who are not at all "interested" in Evolution
Through the avoidance of it for whatever reason
Arrive at the debate of the TOE not fully armed,
And thus fall into the trap of suggesting
That a bunch of lucky chances in a row
Can not amount to anything
And that this proves
An Intelligent Designer.

This is like supposing that
When an organism goes to a casino
And keeps on betting all it has
For billions of years cannot win—
And this is certainly true,
But this is not representative
Of natural selection at all,
For it is not all at once in a row
And so chance is <u>not</u>
The scientific answer to design;
Natural selection is.

Although an organism would
Surely go broke (die)
In this false scenario,
What is also really the case in evolution
Is that there are other organisms
Always taking over (surviving)—
Those that are of a stable platform
(The winnings are already in the bank)
Up to each moment
In between the "gamblings"
On the mutations
(The spin of the wheel
Or the play of the cards
Or the throw of the dice).

This is a simple concept
That still eludes the Creationists
Because they are now so desperate
In their attempts to preserve the invisible.

PURPOSE?

Look to the plants,
They flowering in turn,
And with certain patterns and colors,
So as to gain the full attention
Of the pollinators.

Or even Magnolia
With but plain white flowers
Pollenated by crawling beetles,
For it first grew
Before there were flying insects.

After Darwin sailed to the ends of the earth
On the *Beagle* ship and wrote the *Origin,*
He "retired" to his gardens at Down House,
And continued his many botanical experiments.

These gardens were his engines of war
From which he would lob missiles of evidence
Towards any skeptics—
Descriptions of extraordinary structures
And behaviors in plants
That were very difficult to ascribe
To special Creation or Design,
A mass of evidence
For evolution and natural selection
That was even more overwhelming
Than that presented in the *Origin,*
Through six botanical books
And seventy some papers.

Then, too, there were the 200-plus specimens
That Darwin brought back from the Galapagos,
The single most influential natural history collection
Of live organisms in the history of science.

He questioned the efficacy
Of self-fertilization in plants,
Discovering, by experiments,
That the seeds of those

That were crossbred were heartier
("Hybrid vigour"),
Even that plants had structures
To minimize self-fertilization.

Thus, instead of the same old plant
Just repeating itself again and again,
Which was not the case in nature,
Evolution's change could occur.

His central concern, though,
Was how flowers adapted themselves
To using insects as agents
For their own fertilization.

Thus, again, there were
No direct interpositions of a Creator.

God was not in the details,
But natural selection was,
Acting over millions of years,
Details that were senseless and unintelligible
Except in the light of history and evolution.

What had once just been a pretty picture
Of insects buzzing about brightly colored flowers
Now became a central and essential drama in life,
Full of biological depth and meaning.

The colors and smell of flowers
Had been adapted to insects' senses.

While bees are attracted
To blue and yellow flowers,
They ignore red ones
Because they are red-blind.

On the other hand,
Their ability to see beyond the violet
Is exploited by flowers
That use ultraviolet markings.

Butterflies, with good red vision,
Fertilize red flowers
But may ignore the blue and violet ones.

So, here was even the coevolution
Of plants and insects,
As illuminated by Darwin.

Now, if plants were to reach
The point of reproduction,
They first had to survive, flourish,
And find or create niches in the world,
And so Darwin was, too, was interested
In the devices and adaptations
By which plants survived
And their varied
And sometimes astonishing lifestyles,
Which included sense organs
And motor powers akin to animals.

It is then, the case
That flowers required no Creator,
But were just examples of nature's contrivances,
Wholly intelligible
As products of accidents of selection,
Of tiny incremental changes
Extending over hundreds of millions of years—
And that was the 'meaning' of flowers,
The meaning of all adaptations,
Plant and animal,
The meaning of natural selection.

Darwin had banished meaning from the world—
In the sense of any overall divine meaning of purpose.

There was no design,
No plan, no blueprint in the world,
For natural selection has no direction or aim,
Nor any goal to which it strives.

It spelled an end to Teleological thinking.

THE TELEOLOGICAL,

About beginnings and ends,
They never had a chance,
For there were/are
Neither of those happenings.

The energy of
The fundamental "substance"
Could neither be created
Nor destroyed.

Its arrangements and rearrangements
Came and went / Come and go...
Out of the same old stuff.

ALWAYS AVOID A VOID LIKE THE PLAGUE,

As well as a vacuum, emptiness, nothingness,
Nullity, blankness, empty space, blank space,
Gaps, cavities, chasms abysses,
Gulfs, pits, and black holes
Or else you will be voided invalidated,
Annulled, nullified, negated, quashed,
Canceled, countermanded, repealed,
Revoked, rescinded, retracted, withdrawn,
Reversed, undone, and abolished;
Then you will feel empty, vacant, blank, bare,
Unfilled, unoccupied, and uninhabited,
And very much bereft:
Lacking, wanting, and without,
With nary a snowball's chance in Hell,
Not to mention invalid, null, ineffective,
Nonviable, useless, worthless, and nugatory.

"Now," said the therapist,
"How does that make you feel?"

RECENTLY, AT BOLTZMANNGASSE 3 IN VIENNA

Institut für Quantenoptik und Quanteninformation
(IQOQI)

Does the moon still exist
When we aren't looking at it?

Yes.

LOCALITY AND/OR REALISM (AND 'IT FROM BIT'?)

One (or both) of these assumptions is
Inadequate to describe the physical world;
However, Bell's theorem
Does not say which to abandon.

However, lately it has been confirmed
Even more conclusively
By Ziellinger and associates
That entangled particles do not have
Preexisting properties,
Such as polarization,
That are independent
Of any observation.

So, there goes naive realism.

Now, what about at the classical level?
Well, although there, too,
We transform reality, or I could even say,
Create reality, although it's consistent
Among all individuals,
For we see the same trees
And buildings, for example.

Two particles are called entangled
If they share the same fuzzy quantum state,
Meaning neither of them begins
With definite properties

Such as location or polarization
(Which can be thought of
As a particle's spatial orientation).

Measure the polarization of one photon,
And it randomly adopts a certain value,
Say, horizontal or vertical;

Oddly, the polarization of the other photon
Will always correlate to that of its partner.

Zeilinger, whose group invented
A common tool for entangling polarization,
Likes to illustrate the idea
By imagining a pair of dice
That always land on matching numbers.

Equally mysterious,
The act of measuring one photon's polarization
Immediately forces the second photon
To adopt a complementary value.

This change happens instantaneously,
Even if the photons are across the galaxy;
The light-speed limit obeyed
By the rest of the world
Can take a leap,
For all that quantum physics cares.

I'd like to come to the second freedom:
The freedom of nature.
You said that for example
The velocity or the location of a particle
Are only determined at the moment
Of the measurement, and entirely at random.

I maintain: it is so random
That not even God knows the answer.

For me the concept of "information"
Is at the basis of everything we call "nature".
The moon, the chair, the equation of states,

Anything and everything,
Because we can't talk about anything
Without de facto speaking about the information
We have of these things;
In this sense the information
Is the basic building block of our world.

In your last book you wrote:
"Laws of nature should make no distinction
Between reality and information." Why?

We've learnt in the natural sciences
That the key to understanding can often be found
If we lift certain dividing lines in our minds.

Newton showed that the apple falls to the ground
According to the same laws
That govern the Moon's orbit of the Earth.
And with this he made the old differentiation
Between earthly and heavenly phenomena obsolete.

Darwin showed that there is no dividing line
Between man and animal.

And Einstein lifted the line
Dividing space and time.

But in our heads,
We still draw a dividing line
Between "reality" and "knowledge about reality",
In other words between reality and information.
And you cannot draw this line;
There is no recipe, no process for distinguishing
Between reality and information.

All this thinking and talking about reality
Is about information,
Which is why one should not make a distinction
In the formulation of laws of nature;
Quantum theory, correctly interpreted,
Is information theory.

And can you explain
All these strange quantum phenomena
Conclusively with your information concept?

Not all of them yet, but we're working on it;
With limitation it works excellently.

How?

I imagine that a quantum system
Can carry only a limited amount of information,
Which is sufficient only for a single measurement.

Let's come back to the situation of two particles
Colliding like billiard balls,
And in so doing entering a state of limitation.

In terms of information theory that means
That after the collision the entire information
Is smeared over both particles,
Rather than the individual particles
Carrying the information.

And that means the entire information we have
Pertains to the relationship
Between both particles;
For that reason, by measuring the first particle
I can anticipate the speed of the second,
But the speed of the first particle is entirely random.

Because the information isn't sufficient.

Exactly. Its randomness is ultimately
A consequence of the Finiteness of the information.

Quantum Breakdown

To investigate where quantum mechanics
Breaks down and classical mechanics begins,
The team is investigating
Two weird quantum properties:
Entanglement and superposition.

When two particles become entangled,
They become inextricably intertwined,
So that changing the properties of one
Has an immediate effect
On the properties of its partner

Superposition is another feature
That is peculiar to quantum systems;
Before a quantum object is measured,
It does not have definite characteristics;
Instead, it exists in a superposition
Of multiple mutually contradictory states—
Allowing it to be in two places at once, for example.

Thus, if information is the
Most fundamental notion
In quantum physics,
A very natural understanding of phenomena
Like quantum decoherence
Or quantum teleportation emerges.

And, so, quantum entanglement
Is then nothing else
Than the property of subsystems
Of a composed quantum systems
To carry information jointly,
Independent of space and time;
And the randomness of
Individual quantum events
Is a consequence of
The finiteness of information.

The reduction of the wave packet
Is just a reflection of the fact
That the representation of our information
Has to change whenever the information itself
Changes as a consequence of an observation.

...

A few months ago Zeilinger reported
Implementing a new kind

Of statistical Bell test,
Devised by Leggett,
That pits quantum mechanics
Against a category of theories
In which entangled photons
Have real polarizations
But exchange hidden particles
That travel faster than light.

In principle,
Such faster-than-light theories
Might have perfectly
Mimicked quantum strangeness
And let realism go unmolested.
Not so, according to the experiment:
The results could be explained
Only by quantum unreality.

So what idea replaces realism?
The situation calls to mind
One of Zeilinger's favorite books,
The humorous novel
'The Hitchhiker's Guide to the Galaxy',
By Douglas Adams,
In which a mighty computer
Crunches the meaning of life,
The universe and everything
And spits out the number 42.

So its creators build a bigger computer
To discover the question.

If quantum indeterminacy
Is like the number 42,
Then what idea makes it intelligible?

Zeilinger's guess is information,
Just like a bit, can be 0 or 1;
A measured particle ends up
Either here or there;
But if a particle carries only
That one bit of information,

It will have none left over
To specify its location
Before the measurement.

Unlike Einstein, Zeilinger accepts
That randomness is reality's bedrock.

Still, "I can't believe that quantum mechanics
Is the final word," he says.
"I have a feeling that if we get really deep insight
Into why the world has quantum mechanics"—
Where the 42 comes from—"we might go beyond.
That's what I hope."

"Then, finally, would come understanding."

42

ANTI-MATTERS

The crush on an ant is of the heart
And not of the foot falling part.
But, life, as such,
Steps on us ants,
With so many empty rants,
But, we march ever on
Dwelling in the kingdom
Of strength and toll,
Life ever trying the soul.

Ant-tracks crossed our hearts
When Cupid threw those darts.
Throwing us reeling,
Dealing, feeling, and
Dancing on the ceiling...

If one of us ever needs more space,
One can go to that Saturn place.

TO THE SINGULARITY-TRUTH

As you know, I seek the truth, wherever it leads,
Regardless of any bias of invested interest;
So, I must say that it could be possible
To fill a small point with positive and negative stuff,
Which actually was initially not all the stuff
But a capability to generate
The plus and minus stuff from a balance.

We see that it mostly cancels out,
But for the initial capability,
And, so, as such the larger
Could have come from the small.
So, it was this capability that was "forever",
Not the stuff that is emitted.

However, it is still stuff
And so the painting upon it of a notion of God,
With even umpteen more details added
Does not fly, plus, then one still has to account
For that complex composite Being.

Religion and science must ever
Converge on the causeless.
We then have to ask ourselves how the causeless
Could have any specific certain state of definition
Since nothing preceded it to give it that.

NUMBER REDUCTION

All numbers are great and I love them all;
I wouldn't want to live without any of them,
But, due to the low economy, the "8" digit
Is being dropped, as only a few employ it.

Zero through five make the pyramid known,
And six makes Graham's hexagonal home;
Seven is lucky and nines must end the cost;
So, sorry Prof, eight is downsized and laid off.

AN ALIEN TOUCH ON THE SHOULDER

First of all imagination is
A mental function in your head.

Yes, as all qualia, thoughts, and felt sensations
Are inside the head, even the brightness of light;
Thus, all that we experience is secondary and indirect.

Experiencing a higher dimensional being
Which is not in physical manifestation
Is not an experience of "seeing."
It is a whole body experience
(A felt experience)
Involving activity of the fields
Around the human body
And having to do with what
The ancients called the "astral body",
Which is one's energy body;
It is also matter at a different energy,
Density and frequency level.

Yes, felt sensation appears
In consciousness, as does vision.
The rest of your statement
Is a bias toward what you wish it to represent,
Plus, that it came from 'somewhere else'.

I will document here an experience
You would refer to as an invisible thing....

In 1984, as I lay in bed
And in the middle of
A casual conversation
With my daughter sitting
At the end of my bed,
I felt a force come out
From behind my daughter
And come straight at me.
This was instantaneous!

Notice there are two people involved,
Myself the experience
And my daughter the observer.
The force struck me,
Driving me backwards
Clear across the bed.

My daughter watched as my body
Propelled straight backwards, then rolled over.
She heard my last gasp for breath
And I was seemingly to her...gone, dead,
No response, no breathing,
Even as she shook my body.
I did not feel her shaking me,
(My consciousness was gone.)

I experienced the affects of the force;
My daughter observed the effects
As this force struck my body

You try to propel your body
Backwards across your bed
Without rolling over or using your body
In any way to inch your way backwards.
It is impossible!

Without effort from the body
There is no possibility for motion
Or mobility to go backwards.
Only something striking you
Can propel you backwards.
In my case it was a force
Outside of me or my daughter.

It is not impossible, as muscles can propel,
Especially at a time of going unconscious
When muscle paralysis may not yet be in effect

One cannot go beyond the experience
To say that ETs or angels sent a jolt
From an interdimensional realm.

A full analysis of the jolt
Was not made on the spot;
Speculation is an add-on by imagination,
A bias towards the vested interest
In the source of the happening.

That you went unconscious shows
That consciousness derives
From the physical body and its condition.

Perception is defined as seeing, hearing,
Or becoming aware of
Something through the senses.
We perceive objectively through
Perception or perceptually.
Shermer speaks as though
It is two different things.

Also my experience with
Transformations of consciousness
Had nothing to do with
"Being unable to obtain
A sense of control";
It had to do with a quick wounding
To my established sense of justice,
As the situation I was observing had unfolded
And seemed to hold no justice
For an innocent child and his family.

The 'Mind' can project strange appearances.

There are no conspiracy beliefs;
There are only people with full awareness,
Those in NDE experiences
And also those in the abduction experience.

Those near death would not
Have normal awareness,
But an altered state of it,
As the body is going through dying.

Consciousness is the Soul?

We know that consciousness is of the brain
Because we can introduce
Molecules into brain areas,
By anesthesia, and thus
Turn off consciousness completely.

PROOF FOR '40' BEING THE BASIS OF ALL

Abraham Lincoln,
Biblically named,
Wrote
"Four score and seven years ago"
As part of the Gettysburg address
On the back of an envelop
(Which was the wrong side)
On the way to the battlefield.

Note that "Seven" is almost "eight"
Which is 2 x 4
And that 8 x 5 = 40.

Also, the Holy Trinity
Has been expanded to include Darwin.

4 x TenCommmandmens = 40.

76 trombones + 4 = 80 = 2 x 40.

QED

GOD = THE UNIVERSE?

The universe is not God
Because it is constrained
By the laws of nature,
Which are more restrictive
Than what is logically possible.

THE UNCAUSED, NATURAL STATE OF AFFAIRS

The teleological notion of Who or what
Made or intended the beginning,
Or even an end,
Never even had a ghost of a chance,
For there are neither of those happenings,
Since there can't be any beginning or end
To the uncreated, causeless ground-state—
That which can neither be created nor destroyed.

Here the buck stops,
As there can be no infinite regress
Of causes preceding causes.

No creation equals no Creator.

This ground-state must be the eternal case, also,
Since not anything could have become of Nothing,
This being because Nothing has no properties.

If it was that there had been a lack of anything,
Then this would still be the case.

What about a capability
For pair production of particles
Or even a BiVerse from 'nothing'?

Fine, but those are still secondary
To the capability producing them,
Which is, again, certainly not Nothing.

What about that infinity and eternity
Can never complete?

This is true,
And, so, then,
It must be derived
That only definite forms
Cannot be forever;
But, they were only secondary, anyway,
Being the emissions

That change and move,
And thus became tied to time,
Perhaps this movement of appearances
Even being the definition of time.

It is then that the possibility
From the capability
Of the ground-state
Is still the uncaused eternal,
For what would possibility
Need before itself but itself—
The same thing: possibility.

Nor is the specific amount
Of particle pairs
Or of positive matter
And negative gravitational energy
Any kind of paradox
About Who intended it
To be that amount,
Since this net energy sums to zero;

Nor is its place of origin a paradox,
For that, not being any special place,
Could have been anywhere;

The same with a non-special "when",
For this could be any time.

Nor can God be the fundamental ground-state,
For then ever simpler fundamentals
Would be necessary to compose
A Being who plans, designs, and creates.

So, then, the believers in some great mind
Are looking in the complete wrong direction,
For great complexity comes later on, not first.

Look to the future for even us humans
To become higher and more intelligent beings.

Nor can humans be of any special creation,

For evolution disproves that.

So, all in all,
Any type of God whatsoever
Is disproved on the basis
Of being self-contradictory,
As well as having been cut off
At the very source.

It is even more icing on the cake
That only natural happenings have been seen,
With nothing super-, extra-, or beyond going on.

We could have left it
That believers had nothing
But the invisible
In which to fill in the blanks.
And, incredibly, even many fine details
About a realm that can't even be seen;

But, they preach it as fact and truth,
Which is highly unethical,
This being an outright deception,
For it is only supposed.

And even if believers persist
And soften their stance
By proposing the argument
That "we can never know"
Then it is still the case
That we are completely free to be,
Forever liberated from any notions
Of any puppet strings attached—
Free to make our own meaning in life,
For existence must ever trump essence.

Beyond the nonsense of 'faith',
Which is even defined as a belief
In the non-sense of
The unknowable unseen,
It is still interesting to science
To find out why some humans

Wish to have a predefined
Meaning and purpose,
Especially a divine one
That is really more of
A restriction on freedom,
A lab experiment, even, in some notions.

For this condition, Michael Shermer
And others have many fine answers.

SEXY FLOWERS

Linnaeus, too, was fantastic,
And also a bit playful (or serious)
About the sexuality of flowers,
He sometimes making merry with the idea,
Such as portraying a flower
With nine stamens and one pistil
As a bedchamber in which
A maiden was surrounded by nine lovers.

UPSETTING TO THEOLOGIANS

Linnaeus' research took science on a path
That diverged from what had been taught
By religious authorities; rebuke followed.
The Lutheran Archbishop of Uppsala
Accused him of impiety.

The Catholic Church went further;
Pope Clement XIII banned
The works of Linnaeus
By listing them in the
Index Librorum Prohibitorum in 1758
And also condemned copies to be burned.
Linnaeus was aware of the theological tension
That would be generated
By grouping humans with animals.
(Wiki]

THE BESTED DAEMON

It was the time of positivism,
When Laplace thought
That formulas could tell all,
A time when when so many discoveries
Came about that encyclopedias
Had to be revised every 6 months.

Mary Ann Evans gave up on her first love,
He not finding her body symmetrical.

Her personality made up for this,
Plus her desire to freely fly
From the grasp of determinism—
To transform the purposeless Cosmos
Into her heart's desire.

One was not a puppet;
There were no strings attached;
There was nowhere to attach them to.

She moulded this freedom
Into a love with another
That went beyond all calculation;
They were a poetically scientific couple,
Beneficiaries of the inexactness of formula
And Darwin's facts
Of the arbitrary human narrative,

And, while they knew not the why,
They were freed by the randomness
Of the quantum realm
And other noise in the system,
New brains cells forming
That changed the marble cut
Into a living fluid sculpture.

She thus developed,
Becoming what she was not
In the beginning.

As a novelist,
Mary Ann used her pen name,
George Eliot,
To celebrate this inherent freedom
Long before science discovered it.

GUIDANCE PRINCIPLES?

One would think so from observing the classical realm.

Bohm attempted to speculate a guide wave
And a pilot wave before.

There was shown to be no hidden variables.

It's just that there's nothing beneath
The causeless to guide it.

We can't have causes beneath causes forever:
This makes the unintuitive quantum realm
Somewhat more intuitive.

So, then, a proposed eternal causeless God,
Even more so for its infinite scope,
Would have to have a GREATER DESIGNER
To account for its [perfect] order;
Thus, the eternal causeless
And unordered ground-state must,
Again, be of no design,
And thus not a plan,
A design, a mind, or a God.

Beyond the SuperToe of the causeless,
Of course, then anyone's regular TOE
Of how the particles and forces operate
After 'materialization' can go on as it may unfold.

I don't do much of that nuts and bolts stuff,
For I am a connector of separate ideas,
And also an ultimate kind of guy.

EVOLUTION INTRO

Science matters because it is
The preeminent story of our age,
An epic saga about who we are,
Where we came from,
And where we are going.

— Michael Shermer

Creationism
And its newest reincarnation,
Intelligent Design
Claim that the Creator made
All the forms of the animals directly,
Humans being of a very special creation,
As so indicated in the Bible.

Evolution shows this to be totally untrue,
Although some humans,
Be them never so humble,
Must ever deny it
Since it is emotionally unpalatable—
And might even shy away
From the learning of it,
Yet still going into battle against it,
Though not fully armed.

Evolution unites us
With every living thing on the earth today
And with myriads of creatures long dead,
Giving us the true account of our origins,
Replacing all the myths
That satisfied us for thousands of years.

It shows us our place
In the whole splendid
And extraordinary panoply of life.

This can be deeply frightening to some,
But ineffably thrilling to others.

We are now observing
Species splitting into two,
Finding ever more fossils,
Such as dinosaurs that
Have spouted feathers,
Fish that have grown limbs,
And reptiles turning into mammals—
All demonstrating
The “indelible stamp of our lowly origins”
Of the processes first proposed by Darwin,
Which completely vanquished
The concept of natural theology
Within only a few years by the publication
Of his hundred-page book,
On the Origin of Species,
That turned the mysteries of life’s diversity
From mythology into genuine science.

It is that life on earth evolved gradually,
Beginning with one primitive species
That lived more than 3.4 million years ago,
Which then branched out over time,
Throwing off many new and diverse species.

The mechanism for most (but not all)
Of evolutionary change is natural selection.

Groups like whales and humans
Have evolved rapidly,
While others, like the coelacanth,
The “living fossil”,
Have not changed much
In hundreds of millions of years.

In a creationist explanation of life,
Organisms would not have common ancestors,
But would simply all result from an
Instantaneous creation of forms
Designed anew to fit their environments.

Beyond the live animals and fossils
Used to show similar genes,

And the embryonic phases
Of inherent and older forms
We can now look
At the genes themselves
By sequencing DNA
To construct the
Evolutionary relationships.

The apparent design in nature
Is explained by the
Purely materialistic process
That doesn't require creation
Or guidance by any supernatural forces,
For individuals of a species
Vary genetically in their ability
To survive and reproduce
In their environment.

There's a real difference
In what one would expect to see
If organisms were consciously designed
Rather than if they evolved by natural selection;

Natural selection is not a master engineer,
But only a tinkerer.
It doesn't produce the absolute perfection
Achievable by a designer starting from scratch,
But merely the best it can do
With what it has to work with,
Keeping the organism's structure
Habitable all the while.

It produces the fitter, not the fittest.

What conquers our ignorance is research—
Not giving up and attributing
Our ignorance to the miraculous work of a Creator.

We are apes descended for other apes,
And our closest cousin is the chimpanzee,
Whose ancestors diverged from our own
Several million years ago in Africa.

We don't stand apart from the rest of nature,
However disconcerting that may feel.

There is evidence from many areas—
The fossil record, biogeography,
Embryology, vestigial structures,
Sub-optimal design, DNA, and so on—
All of that evidence showing,
Without a scintilla of doubt,
That organisms have evolved.

We don't find mammals in Precambrian rocks
Or humans in the same layer with dinosaurs.

Despite a million chances to be wrong.
Evolution always comes out right.

Special creation does not;
It goes wrong in every area,
And not just in the area of evolution.

Many people can't or won't accept evolution,
For it raises profound questions
Of purpose, morality, and meaning
That are emotionally hard to face
As the consequences of our
Evolution from apes—
And so they can't, then,
Deal fully armed with what we are.

Like all species,
Human beings evolved
From the working of
Blind purposeless forces
Over eons of time.

That is the real story of our origins:
Naturalistic materialism.

The *Origin of the Species* might now just as well
Supplant the Bible n defining the wonders of nature.

THE AGES OF THE SPECIES

Simple photosynthetic bacteria,
The beginnings of us and all,
Appear in sediments
About 3.5 billion years ago,
Only about a billion years
After the planet was formed.

These single cells were all
That occupied the earth
For the next 2 billion years,
After which we see
The first simple "eukaryotes":
Organisms having true cells
With nuclei and chromosomes.

Then, around 600 million years ago,
A whole gamut of relatively simple
But multicelled organisms arise,
Including worms, jellyfish, and sponges.

These groups diversify over
The next several million years,
Along with terrestrial plants and tetrapods
Appearing about 400 million years ago,
Tetrapods being four-legged animals,
The earliest of which were
Lobe-finned fish.

50 million years later
We find the first true amphibians,
And, after another 40 million years
Reptiles come along.

The first mammals show up about
250 years ago,
Arising, as predicted,
From reptilian ancestors.

Mammals,
Along with insects and land plants,

Continue to diversify
As we approach the shallowest rocks,
The fossils increasingly resembling living species.

Humans are newcomers on the scene—
Our lineage branches off from that
Of other primates only about
7 million years ago,
But a sliver of evolutionary time.

If the entire course of evolution
Were compressed into a single year,
We would not see the first human ancestors
Until 6 AM on December 31.

The golden age of Greece,
About 400 BC,
Would occur just thirty seconds
Before midnight.

Happy New Year to all!

FROM FISH TO AMPHIBIANS!

Found at the ends of the earth.

In 2004, a transitional form
Between fish and amphibians
Was discovered—
The fossil species of *Titaalik roseae*,
Which tells us how vertebrates
Came to live on land;
Yet another stunning vindication
Of the theory of evolution.

Until about 300 million years ago,
The only vertebrates were fish;
But, 30 million years later,
We find the early tetrapods:
Four-footed vertebrates
That walked on land.

They were like modern amphibians
In that they had
Flat heads and bodies, a distinct neck,
And well-developed leg and limb girdles;
Yet, they also show strong links
With earlier fishes,
Particularly the group known as
"Lobe-finned fishes",
So called because of their large bony fins
That enabled them to prop themselves up
On the bottom of shallow lakes or streams.

The fishlike structures of early tetrapods
Include scales, limb bones, and head bones.

Neil Shubin had spent years
Studying the evolution of limbs from fins.

Since there were lobe-finned fishes
But no terrestrial vertebrates
390 million years ago,
And clearly terrestrial vertebrates
360 million years ago,
Then the fossils of the transitional forms
Should be found in strata
Around 375 million years old.

Moreover, since late lobe-finned fish
And early amphibians both lived in fresh water,
The rocks would have to be from freshwater
Rather than marine sediments.

Shubin and his colleagues zeroed in
On a paleontologically unexplored region
Of the Canadian Arctic:
Ellemere Island, which sits in the Arctic Ocean
North of Canada.

Five years of fruitless searching passed
Before they finally hit pay dirt:
Skeletons stacked on atop another
In sedimentary rock from ancient streams.

In honor of the local Inuit people
And the donor who helped fund the expedition,
The fossil was named *Titaalik roseae*,
"Titaalik" meaning "large freshwater fish" in Inuit
And "rosae" being a cryptic reference
To the anonymous donor.

It had gills, scales, and fins for the water,
But eyes and nostrils on the top of the skull
Rather than on the sides,
It's head flattened like that of a salamander;
It could peer, and probably breathe,
Above the surface.

The fins had become more robust,
Allowing the animal to flex itself upward
To help survey its surroundings,
And, like the early amphibians,
Titaalik had a neck.
Fish don't have necks—
Their inner skull
Joins directly to their shoulders.

Titaalik also had two novel traits
That would help its descendants invade the land,
The first being a set of sturdy ribs
That helped the animal pump air into its lungs
And move oxygen from its gills,
It being able to breath both ways,
And, second,
Instead of the tiny bones
In the fins of lobe-finned fish,
Titaalik had fewer
And sturdier bones in the limbs—
Bones similar in number and position
To those of every land creature
That came later, including us;

In fact, its limbs are best described
As part fin, part leg.

Now, somewhere in freshwater sediments

About 380 million years old,
We'll find a very early land dweller
Wit reduced gills and limbs a bit sturdier
Than those of *Titaalik!*

BIRDS

Did birds arrive, flying,
Out of thin air?

No, and the short story is
That gliding was the first step,
As seen in the fossils,
Then the full becoming of birds
From their common ancestor
With the dinosaurs, the *theropods:*
Agile, carnivorous dinosaurs
That walked on two legs.

And there the transitional are
In the predicted strata,
Written into the rocks of ages,
145 million years ago,
One being *Archaeopteryx lihographica,*
Discovered in a limestone quarry
In Germany in 1860,
Darwin's time.

EMBRYONIC FORMS

In their development from embryos.
Many species go through
Contortions of forms that
Are bizarre organs
And other features that appear,
And then change dramatically
Or even disappear completely before birth.

Mammals like ourselves
Even produce a yolk sac—

One that is vestigial and yokeless,
A large fluid-filled balloon
Attached to the fetal gut,
Harking back to the egg-laying days
Of our reptilian ancestors;
However, in the second month
Of human pregnancy,
It detaches from the embryo.

Our blood vessels go through
Especially strange embryonic contortions.

In fish and sharks,
The embryonic pattern of vessels
Develops without much change
Into the adult system;
However, as other vertebrates develop,
The vessels move around,
And some of them even disappear.

Mammals like ourselves
Are left with only three main vessels
From the original six,
These changes curiously resembling
An evolutionary sequence,
The first being that
Of embryonic reptiles,
To which more twists and turns are added.

This "recapitulation" of an evolutionary sequence
Is seen as well in the development of other organs:
The human embryo actually forms
Three different types of kidneys,
One after the other,
The first two discarded
(Resembling those of jawless fish
And reptiles, respectively)
Before our final kidney appears.

There is also a fetal coat of hair
That then greatly diminishes.

THE ASCENT OF HUMANS

There was no act of special creation.

Evolving from those before it,
Homo habilis, the "handy man"
Who first used tools,
Appears about 2.5 million years ago,
And may or not be a direct ancestor
Of *Homo Sapiens,*
But *habilis* does show changes
Toward a more human-like condition,
Such as reduced back teeth
And a brain larger than that
Of the *austrapithecines*
And it also shows parts of the left lobe
Corresponding to Brocca's area
And Wernicke's area,
Parts associated with speech
And comprehension.

Home habilis, who may have coexisted—
In time if not in space—
With other hominins,
Including some that went extinct,
Such as the "robust,
(As opposed to graceful hominins)
Being *Paranthropus australopithecus,*
Robustus, and *aethiopicus,*
As well as with three species of *Homo:*
Homo egaster, rudolfensis, and *erectus.*

Homo erectus ("upright man)
Was the first hominid to leave Africa,
Its remains being found in China
As "Peking man",
Indonesia ("Java man"),
Europe, and the Middle East.

By the time of this Diaspora,
The brain size of *erectus*
Ws nearly equal to that of modern humans,

Although they still had a flattened
Chinless face—the chin being a hallmark
Of modern *Homo Sapiens.*

Their tools were complex.
And they had control of fire—
A momentous event.

Homo erectus was around
For about 1.5 millions years,
Disappearing from the fossil record
About 300,000 years ago.

It may have left two famous descendants,
Homo heidebergenis,
Know as "old archaic Homo Sapiens",
Who first appears .5 million years ago,
With a mixture of *H. Sapiens*
And *H. erectus* features,
And *Homo neanderthalensis,*
Appearing 380,000 years ago.

What ever happened to *H. erectus,*
Every *H. erectus* population
Suddenly vanishing,
About 300,000 years ago,
Being replaced with
"Anatomically modern"
Homo Sapiens?

And what of Neanderthals,
Who hung on awhile longer,
After finding their last redoubt
In caves overlooking
The Strait of Gibraltar?

And, in the future,
What of us?

THE UNCAUSED GROUND-STATE

It's more that any given ground-state
Would have no specific or certain definition,
There being nothing prior
To give it any special definition.

It could even be complete chaos, whatever that is.
However, after it bloomed into some arrangement
That continued on, then there should be
Some cause and effect that carries on.

Science and religion must ever seem
To converge on the causeless,
Although the first may wish
To claim perfect order there.

There is a difference between saying
The origin of the universe has an explanation
And saying it has a cause.

The quantum fluctuation process
Here established a set of conditions
Which then permit the tunneling
Of a cosmology.

These are what the initial
And boundary conditions of the universe,
Which are set by noncausal quantum processes.

So in this setting
There does not exist a set of initial conditions
Which define a "first cause";
Rather, quantum fluctuations in a vacuum
Resulted in a topological configuration of space,
Or 10 dimensional superspace,
Which by stochastic means
Contained field theoretic information
Which inflated into a cosmology.

NOBODY NOWHERE

The ToeQuest member,
Named Nobody Nowhere,
Who we think has died
Since he contributed every day
And suddenly stopped,
Has a thread of 'The Theory of Nothing'
And another one 'Nobody Nowhere',
Thought that Nothing
Could differentiate itself into Everything.

But you see then that his "nothing"
Was not an absolute Nothing,
For it had this
Inherent potential/capability
To divide itself.

Anyway, he thought that Everything
Already happened, long ago,
And is all over and done with,
But that it takes time to propagate
At the speed of light
And so it is still playing out for us.

He also had gravity, a negative force,
Erasing everything,
Just after it briefly happened,
Via light, a positive force.

THE ECONOMY

Money is real but stock is not;
The riches there went to rot.

WHAT IS A DREAM?

Is a dream defined as an alternate reality
In which we manipulate our own chaotic existence?
The imagination of each and every individual
Is the foundation on which dreams are based,
And based solely upon the element
That is each individual,
Which is the prospect of personality.

It is, for we remain inside ourselves
In the solitude of our own consciousness
That observes and witnesses what we have become.

While awake, we have access to more information,
Especially the internet, but, when asleep
And dreaming we are ever more in our own isolation.

Being that the basis of a dream
Is depicted upon an elementary soul,
Or conscious/ subconscious foundation,
The soul and reality of a
Dream maintain a balance,
Thus resulting in the dream itself.
This being said,
What is the absolute definition of a dream?

I might only suggest that the inner mind,
Absent any real inputs from wide awake thoughts
And of sense from the outside,
Still tries to make sense of its inputs,
At its deeper interface, those perhaps being noise,
Static, and some subconscious wanderings.

Thus, dreams would not just report noise
And static as coming in, showing a picture of that,
For it wouldn't even know that,
And so it would take off
On some of the jiggling
Of its up-stage inputs
About what might be happening

And make scenes that then take off
From what's left awake of the imagination,
That being a fair presentation,
Although backgrounds
Often change at random,
Stuff doesn't function right,
That gravity can go away, and, for me,
That my car is never where I left it.
To fully comprehend a dream,
You must understand that reality is a dream,
And vice-versa
What depicts the fate of, "reality",
Is the concept of nature,
Being that the general concept
Of nature is that it is our origin,
From where we existed
Through its imagination.

It is a dream, in a way,
But the remainder of your theory
Is a personification of nature
As being intentional by having imagination.

Since it is nature's imagination
From which we exist,
There must exist a contradiction
Of a conscious/ subconscious
Nature within nature itself,
Meaning that the, "reality",
Of a conscious/ subconscious nature
Is actually an equal balance of nature's reality
And the concept of its dream.

Shall we rather say that nature just does what it does,
Given its blind, purposeless ways shown for sure
By evolution via natural selection?

Therefore, the contradiction creates
An endless median of a configuration
Consisting of an equal balance

Between the polar opposites, or extremes,
Which means that any contradiction,
Such as a dream vs. reality is actually one in the same,
Being that one could not live without the other.

The polar opposites of positive and negative,
Such as we see in quantum pair production,
Some of which have become separated
Into the rather enduringly real,
Allow for the secondary effect
Of all the many possible configurations.

Further, imagine the reality of a dream,
Whether in a sleep state or being awake,
Considering they are one in the same,
And in your imagination you imagine a dream,
And in that dream you imagine another dream,
And in that dream you imagine yet another dream,
To the point of singularity.

It has happened to me that I thought I awoke,
But was still in the dream,
Or then even went to sleep in my dream
And had another 'dream' dream.
Here, I was untouchable by all external reality,
Freer than free could ever be.

Since we are the prospect of natures imagination,
Every choice we make, that is made for us,
Is yet another dream outside
Of the previous dream experienced
(In my theory, mathematical chaos
Would dissect time infinitely)
And each moment would be the same,
But also different, infinite but non existent.
The laws of physics would cease to exist
In a pure, infinite state,
A point relative to its own instinct,
Irrelevant and separate of knowledge.

Nature still has no imagination,
And there is no true infinity or eternity,
For those terms mean
"That which can never be attained".

I believe infinite is a level of acceptance,
Or belief, somewhat letting go of knowledge itself
And entering a realm of instinctual understanding.
For example, I believe an inanimate object
Is purely instinctual, and resides in a realm
Of infinite, because it does exist,
But is fully directed by its surroundings
Although itself and the environment
Share a common boundary.

No actual infinite,
But there are a heck of a lot
Of finite things going on.

Because there exists infinite natural decisions
Within each moment of singularity,
And each action contained within the singularity
Exists as a separate dream,
Depending on how you perceive it,
Like where each dream begins and ends,
Endless and infinite dreams
Exist within each singularity.

It is potentially endless,
Within the limits of form while awake,
But, more endless within a night dream.

This is a possibility due to the fact that time,
Which determines where something begins and ends,
Can be infinitely spliced to smaller intervals
Through mathematical chaos, and therefore,
Time stops and ceases to exist
The moment singularity is achieved.

This is the potential infinity
Or eternity of math,
But not an actual one.

Therefore, if each separate dream begins and ends
On the foundation of singularity,
It is never ending but also never beginning,
Because "infinite" is "all of time"
And "all of time" extends forever
Like a sphere surrounding the edges of existence
But also somehow containing itself continually,
Which is the only way it would be "infinite".

The "timeless" simply exists, rather than not,
For it must do so,
As nothing can become of Nothing
And there cannot be
An infinite regress of causes.

We break barriers,
Only to discover each and every boundary
Is encased by another boundary.
Is this some kind of sick joke?

No, for no one played it.
There is only the causeless
Ground-state as the last boundary,
A necessarily arbitrary state
From which all came,
Whenever, wherever, and however,
For none of these particulars were "special".

One could take it
As seeming like a Cosmic joke,
In that there is no specific purpose.
Existence precedes essence
In importance by a long-shot.

THE DREAMER

The shadowy and imaginary figure of
The Keeper of the Kinds lurks unseen
In the center of the universe.

Once he was the sun shining on all mankind,
Giving them life in their time,
Or perhaps he had only dreamed this
On some hazy day long ago.

Days of lightning thundered into nights
As the same day rebroke again and again,
The cycle of the seasons ever going 'round.

He had long since forgotten his birth
Through fire and time,
Having no earliest memory, anyway.

Silently he walked,
Even though there was no one to
Hear the footsteps of a lone man
Walking nowhere through empty space.

He walked on and on—
But he could never escape
From the center of the Universe,
For it had no boundary, no circumference,
And therefore its center
Was everywhere and nowhere.
Finally, he sat down.

In his mind's eye at
The center of a Universe
Receding in all directions,
The Keeper of the Kinds turned
Ever so slowly in his chair
And stared out the window into Humanland.

He cared little for what he saw
Since he'd seen it all before.
He cared even less about me or you.

Most of the time he cared only for order,
And rarely for naught except
On those hyper days when he wondered
I fleas had fleas or if he might ever
Become his own Keeper's Keeper.

Well, this was one of those days,
And so on this day the Entropy Devil
Was kinged for a time.

Henry Humpersnickle, one of the kind,
Was indeed wary of being caught up
In the scheme of things,
So he stumbled onto an escape from reality.

After Henry went to sleep
He dreamed that he had awakened,
But, at first, upon actually awakening,
He didn't even remember it;
But, that was good,
For then neither did the Keeper
Take much note of it either.

Thence, Henry awoke,
In dream only,
In a strange world,
But he thought that it was real.

All of these events had almost happened before,
But were unique since one grain of sand
Had shifted ever so slightly
(By the length of a blue light wavelength).

The Keeper, an eternal determinist, was not upset,
For he knew that this might happen someday,
As sure as he knew that the entire contents
Of an encyclopedia might be represented somewhere
In the non repeating expansion of pi (3.1416...)
That went on forever, all of which, of course,
He tried to hold within the extent of his mind,
But, of course, could not.

However, lately there was talk from his own Keeper
That potential infinities need not be exhaustive.
Nevertheless, he could never know everything,
And didn't care to anymore,
for only his own Keeper could unlock
Life's two Yin/Yang boxes,
Each of which contained the other's key.

Meanwhile, a Bishop at Queen's Knight 10**11**9
Had attacked the pawn at King's Bishop 5**5**6,
Diverting the attention of the Keeper
And sending illusions of ripples
Through Henry's world line.

Although it was still questionable
As to whether all events
Must eventually happen in a world of illusions,
Henry had already made the question academic,
For Henry had now dreamt of dreaming,
And, what's more,
He became very much aware of it
And was quite lucid;
Thus, the Keeper's grip on him loosened,
And Henry's ripples became a much smoother.

Soon there would be no sign
That Henry's pebble had even slipped
Beneath the surface.

Indeed, it could no longer even be determined
If the pawn was still under attack,
Or even who this new Henry was,
For there was no one around
To answer the question.

The Keeper did not miss
Henry's new incarnation,
For the elements of his Universe
Still constituted a tautology
On Nature's thumbnail.

Ice winds filled Henry's vacuum,
And, as he dreamt of dreaming and awakening,
The fates of his chances answered
To none other than the chances of his fate.

As his own Keeper,
Henry kept to himself.
Being alone, as a being alone,
Henry no longer bothered
With keeping track
Of time or movement since
This was quite impossible
With no one around.

It was all he could do
To remember the day
That the monsters came.

THE POLAR EXPRESS

At the pole of the north,
All the winds blew south,
Then souther and souther south,
Until all the winds blew north.

The cold from the frigid North formed a dome
That was quickly turning everything to stone.

The clunker car turning over sounded a groan,
And she was really frozen all the way home,
As there was little meat and fat on her tiny bones;

But her brain was thinking meat,
(Forget the soul; it's cold.)
Ever wishing for the heat.

A CONTAINER OR A PORTAL?

It's a container, where the buck stops,
Although just a possible one,
But noted by the example of the quantum foam
Producing plus and minus pairs of particles,
Coming out of this container-portal, so to speak.

My other TOE is that a certain amount of energy,
Like that of FS, was always around,
Going through rearrangements,
Even portals, or even cyclical universes.

But two paradoxes then present themselves:
1) Real and defined stuff
Could not have been eternal up to now,
For eternities and infinities
Can never complete, by definition;
However, if timeless,
Then it simply is, rather than not,
But I don't know if that gets us out of it.

2) How could there be a certain, specific amount
Of real stuff around without that amount
Having ever been defined in the no-first-place.
Not to mention the where and when
And its properties other than the amount.

Thank you for a very distinct answer, Austin
What surprises me, though,
Is that you use the words eternities and infinities
As if they are actual entities.
These are conceptual entities only,
That all by themselves contain nothing at all.
So, why use the empty position
To pose a question with these conceptual entities?

Same goes for time;
There is nothing in time unless there is matter.
Why place the conceptual entity Time
In the empty position and then stating
There is something actual there?

I am using the words,
But in their defined context
Of "that which can never be attained" (completed)
To show that they are impossible, as entities,
Although we can still employ them potentially,
Such as "heading out [toward] forever"
Or that "numbers keep going on and on".
So, it is then that whatever stuff is tied to time
Could not have already been around forever.
This is resolved if time was created along with it,
Whatever time is, but time could just be change.

For your second point the answer
Is in my opinion very simple.
If you accept that energy was always around,
Then the universe is the materialization of energy,
The sequel so to speak of the original energetic state.

Since energy must have overcome its own restrictions
To become materialized energy,
The result is mainly self-based
And the result will therefore
Show a lot of symmetry.

Since energy could overcome its restrictions
Only because it got into a conflict with itself,
It overcame its restrictions by pushing the conflict
All the way up to the level of the restrictions.
Then, a single step into the unknown
By the smallest part imaginable was all it took
To have this giant pyramid scheme come undone
Onto the new reality where materialization
Became the essential entity.

I still do not know how to interpret the term FS.
Do you mean to say that energy always was?
That matter is just the expression of energy
In a peculiar state (a certain stuck state)?

But why use the term FS then,
Instead of saying that energy always was?

I asked Lloyd and Pat,
But I still did not understand their explanation.
Possibly, they proclaimed an entity,
While only trying to declare a conceptual state?
I can use your help to understand this term

The symmetry would be
The positives and negatives
That tend to cancel out,
But only if they all get near each other,
Which, evidently, they all didn't.

I am loosely using energy
And FS (Lloyd's) interchangeably;
It would be that FS always was,
Since that's what would have the energy,
Or that waves whizzing all over
Could get dense enough to form "substance,"
But that doesn't matter much,
Or, rather, it is matter.

So, I guess I'll conclude that
It is only the secondary stuff
That is tied to time and that
The special amount of that stuff
Is of no concern since it can vary
And always cancels out, potentially.

To me, the bottom of the pyramid
Would have to be the "uncaused",
Since nothing can become of Nothing
And that no state could be
Prior to the "uncaused";
I'm saying that whatever
Gave rise to the universe
Was itself uncaused.

And that is the TOE found!

A BRIEF HISTORY OF ALL OF HISTORY

For all time
Vacuum fluctuations
waver in and out of existence
since nonexistence cannot be.

Our Universe

1E-43 seconds
Planck era.

Cyclical compactfication
or a
Vacuum fluctuation eruption

1E-36 seconds
GUT transition.

Strong force separates
from the Electroweak force.

1E-36 seconds
Inflation begins.

Slow rolling scalar field
generates negative pressure
causing exponential expansion
of space time.

Doubling time: 1E-36 seconds.

Vacuum energy density: 1E73 tons/cm^3.

Quantum fluctuations lock in
nearly scale invariant 1E-5
variation in energy density.

1E-34 seconds
Inflation ends.

Decay of scalar inflaton field
causing reheating.

Is this the let there be light moment?
No, photons don't exist yet,
but other massless vector quanta
like left and right weak
and B-L particle may exist.
Things are not well known about this era.

1E-34 to 1E -8 seconds
Quark era.

Quark gluon plasma.

Quarks and super particles
dominant matter content.

1E-17 to 1E-15 seconds
SUSY breaking.

Super partners acquire mass
with the LSP expected to have
a mass of about 10 Tev.

(In induced Gravity model,
this is where mass energy
first generates the
induced gravity field,
Gravity is born.)

1E-10 seconds
Electroweak transition.

The Electroweak force,
under the action of the Higgs mechanism
breaks symmetry.

The photon is born.
Standard model particles get mass.

1E-5 seconds
Quark confinement.

The QCD vacuum becomes superconducting
to color magnetic current.

Quarks and Gluons are confined.

1E -5 to 1 E-4 seconds
Hadron era.

Hadrons are formed:
protons, neutrons, pions etc.

1E -4 seconds
Hadron annihilation.

A brief period of proton/anti proton
and neutron/anti neutron annihilation.

A slight favoring of matter over anti matter,
possibly locked in by CP violation
at Reheating causes some protons
and neutrons to survive.

1E-4 to 10 seconds
Lepton era.

Following Hadron annihilation
Leptons are the dominant energy density.

1 second
Neutrino decoupling.

Mass energy falls low enough to free neutrinos,
creating the neutrino cosmic background.

10 seconds
Electron annihilation.

Electrons and positrons annihilate,
leaving a tiny fraction of electrons remaining.
At this point the total number
of electrons equals the total number of protons.

10 seconds to 57 thousand years
Radiation era.

Photons created from
the annihilation of matter and anti matter
dominate the energy density of Universe

1- 5 minutes
Nucleosynthesis

Fusion of protons create helium,
deuterium and trace amounts of Lithium.

57 thousand years
Matter/radiation equality.

The radiation density
(photon and neutrino)
and matter density
(dark and atomic)
are equal.

This is because radiation density
falls more quickly due to the stretching
of the relativistic particles wavelengths.

Dark matter clumps into structures.

Atomic matter begins oscillation
due to the battle between gravity
and photon pressure
generating acoustic oscillations.

The first sounds of the new Universe

380 thousand years
Recombination.

The temperature falls low enough
to allow atoms to form; photons decouple.
The CMB is born, locking in its structure.

The story of the earliest times in the Universe.

5 to 200 million years
The dark age.

Photons fall into the infra red energy range;
the Universe goes dark.

The atomic gas continues to fall toward
the dark matter clumps
which grow more pronounced.

Near 100 Million years
the densest clumps
halt their expansion and begin collapsing.

By 200 Million years
the first mini halos
form and within these
the atomic cloud cools and collapses
to make the very first stars
whose light brings to an end the dark era.

200 million years
First stars.

The first stars are very massive and short lived.
They die in violent Super Nova explosions
filling the cosmos with the building
blocks of planets and the elements
needed for life.

200 to 800 million years
Epoch of ionization.

The radiation from the stars
and possibly the first quasars,
ionizes much of the remaining
neutral hydrogen and helium.

A thin mist returns
and partly obscures the CMB.
(Future Low Frequency Radio Telescopes
may soon be able to see the epoch of ionization)

1 to 2 billion years
Infant galaxies.

Star groups merge,
forming the very first galaxies.

There are frequent collisions of galaxies,
high star birth rates
and high supernova rates.

Heavy element production
changes the pattern of star formation,
making them lower mass,
less luminous and longer lived,
like those of today.

The stage is set for the emergence of life;
the Cosmos will soon
have eyes to see and minds to think.

2 to 3 billion years
Star birth and quasar peak.

In the dense environment
of frequent galaxy collisions
the star birth rate
reaches it maximum,
as does the forming and feeding
of supermassive black holes.

6 billion years
First rich galaxy clusters.

Enough time has elapsed for the densest
regions to stop expanding and form clusters.

7 billion years
Deceleration /acceleration.

The effects of Dark energy kick in.
The Universe once again begins
to accelerate its expansion rate,
but at a much more gentle rate.

8 billion years
First modern spiral galaxies.

Although some elliptical galaxies
form in the first billion years,
classic spiral galaxies aren't seen
until about 5 Billion years ago.

9 billion years
Matter / dark energy equality.

At this time the falling density
of matter (dark and atomic)
become equal to that of dark energy.

9.1 billion years
Sun and Earth form.

The solar system forms
in the outer disk of the milky way.

The stage is set for the emergence
of humankind in the Cosmos.

13.7 billion years.
Present time.

Human civilization reaches its peak
and begins heading into decline
and eventual extinction
due to over population,
resource depletion,
and environmental destruction
which generates conflict
as Human nation states fight
for ever dwindling resources.

Hopefully Humankind is not typical
and intelligent life elsewhere
solves the problem of balancing
intelligent life needs
with available resources
by developing communitarian
economic social structures.

April 16, 2010
Austin organizes all of history.

16 to 17 billion years
The Milky way collides
with the Andromeda galaxy.

Somewhere within this time
the Sun enters into its Red Giant phase,
vaporizing the earth.

Humankind extinct for over 4 Billion years
is not around to witness this event
though possibly a new intelligent species
which emerged after
the extinction of Humankind might be.

It will be a very sad time for them
unless their technology includes
very advanced space flight.

20 billion years
Growth of Structures cease.

Expansion due to Dark energy
empties each casual patch of the Cosmos.

The story of our Universe draws to a close.

100 billion years
What remains of the Milky way
is alone in its causal patch of the Universe.

1000 billion years
Last stars die.

The Universe is empty and dark.

However, stirring in the vacuum
of space time itself
are the ever present
vacuum fluctuations.

One small patch quite by chance
fluctuates sufficiently to create a
volume of false vacuum which cuts off
from its mother Universe
by negative pressure,
explodes into a new Universe
creating new space time and future hope
for the emergence of intelligent life in the Cosmos.

Everything starts over.

— THE END —

www.ingramcontent.com/pod-product-compliance
Lightning Source LLC
Chambersburg PA
CBHW071401170526
45165CB00001B/141
9781451560114